Coatings for Advanced Devices

Coatings for Advanced Devices

Guest Editor

Xu Long

MDPI

Basel • Beijing • Wuhan • Barcelona • Belgrade • Novi Sad • Cluj • Manchester

Guest Editor
Xu Long
School of Mechanics, Civil
Engineering and Architecture
Northwestern Polytechnical
University
Xi'an
China

Editorial Office
MDPI AG
Grosspeteranlage 5
4052 Basel, Switzerland

This is a reprint of the Special Issue, published open access by the journal *Coatings* (ISSN 2079-6412), freely accessible at: https://www.mdpi.com/journal/coatings/special_issues/Adv_Device.

For citation purposes, cite each article independently as indicated on the article page online and as indicated below:

Lastname, A.A.; Lastname, B.B. Article Title. *Journal Name* **Year**, *Volume Number*, Page Range.

ISBN 978-3-7258-3765-6 (Hbk)
ISBN 978-3-7258-3766-3 (PDF)
https://doi.org/10.3390/books978-3-7258-3766-3

Contents

About the Editor

Xu Long

Professor Xu Long received his Bachelor's Degree in Engineering Mechanics from Tongji University in 2006, Master's Degree in Solid Mechanics from the Chinese Academy of Sciences in 2009, and Ph.D. in Structural Mechanics from Nanyang Technological University in 2013. After working as a Senior FEM Engineer in INTECSEA, WorleyParsons, Singapore, he joined Northwestern Polytechnical University, China in 2015. Long's research interests are in the fields of the mechanics of electronic packaging and protective material and structures, including multi-field multi-scale material constitutive and damage model life prediction and reliability analysis. In recent years, his research group has published more than 150 peer-reviewed journal papers in the areas of mechanics and mechanical materials, including more than 90 SCI-indexed papers for which he was the first author or corresponding author. He has published three books as the first or only author. He currently serves as the Associate Editor of *Computer Modeling in Engineering & Sciences* and *Journal of Polymer Materials* and an Editorial Board Member of *International Journal of AI for Materials and Design*. He has also been invited to be the Guest Editor-in-Chief of a number of journals. He was selected as Young Top Talent of the "Ten Thousand Talents Plan" China, a IAAM Fellow, and a IEEE Senior Member. He has won over 20 national and international academic and industrial awards. He has also actively served as the Technical Program Committee or Session Chair for more than 20 international conferences, including the International Conference on Electronic Packaging Technology and the International Conference on Computational & Experimental Engineering and Sciences.

Preface

This Special Issue reprint, Coatings for Advanced Devices, highlights recent progress in the development and application of functional coating technologies that are essential to the performance and reliability of cutting-edge electronic and multifunctional systems. From microelectronics and optoelectronics to flexible displays and wearable healthcare devices, coatings play a critical role in enabling advanced functionalities and protecting devices that operate under increasingly demanding environmental and mechanical conditions.

The aim and purpose of this reprint are to provide a comprehensive overview of cutting-edge research in coating materials, thin-film processing, and interfacial engineering, as well as to promote the exchange of ideas between academic researchers and industrial engineers. It brings together contributions that address a wide range of challenges, including the design of adaptive and multifunctional coatings, enhancement of interfacial adhesion, performance in extreme environments, and integration into scalable device fabrication processes. By linking fundamental insights with technological implementation, this Special Issue pursues a deeper understanding of how surface and interface engineering can drive the advancement of next-generation devices.

The reprint is intended for scientists, engineers, and graduate students working in the fields of electronic packaging, solid mechanics, materials science, and electronic manufacturing. We hope that the works presented will not only serve as a valuable reference but also inspire continued innovation across disciplines that intersect at the interface of materials and devices.

I would like to extend my sincere gratitude to all contributing authors for their high-quality submissions, and to the reviewers for their insightful and constructive feedback, which significantly enhanced the quality of each manuscript. I also wish to express my appreciation to the editorial team, especially Ms. Sunny Yang, for their professional support and dedicated assistance throughout the publication process. This work was supported by the National Natural Science Foundation of China (Nos. 52475166, 52175148) and the Regional Collaboration Project of Shanxi Province (No. 2022040411101044), whose contributions are gratefully acknowledged.

Xu Long
Guest Editor

coatings

Editorial

Coatings for Advanced Devices

Xu Long

School of Mechanics, Civil Engineering and Architecture, Northwestern Polytechnical University, Xi'an 710072, China; xulong@nwpu.edu.cn

The continuous evolution of coating materials is critically important to driving innovative technologies in various domains, especially advanced electrical devices. This is because most current devices and systems operate under ever-complicated and harsh conditions, which makes the demand for coatings with enhanced mechanical, thermal, and functional properties increasingly high. Recent research has focused on the development of new coating compositions, advanced deposition techniques, and innovative material designs to improve interfacial adhesion and mitigate thermal stresses, enhance corrosion resistance, and optimize optical performance.

This Special Issue combines 10 papers and brings together a collection of studies that address critical challenges in innovative research and applications based on coating technologies, ranging from electronic packaging and high-temperature environments to sustainable polymer coatings and optical protection. Leading scientists and engineers discuss the challenges and breakthroughs that arise within coating technologies that emphasize how high-end coating technologies shape the next generation of high-performance electronic devices. The contributions put together related knowledge from the areas of computational modeling and experimental characterization that outline the constantly evolving landscape of coating materials for their potential in enabling next-generation technologies and defining the future of the development of sophisticated devices, attaining unprecedented levels of reliability, efficiency, and versatility. We invite the reader to learn about the transformative effect of pioneering thin film/substrate systems in advanced devices that are pushing the boundaries of electronics, energy, and beyond, ushering in an era wherein technology and sustainability will thrive in harmony.

Long et al. [1] proposed a method for determining the elastoplastic properties of thin films by indentation using a dimensionless analysis. This is a key tool in material characterization for electronic packaging and other microelectronics applications. The investigation by Su et al. [2] extended this understanding by proposing fiber-reinforced composites that can be used to tune the CTE for multilayered coatings—an effective mitigation approach against thermal stresses in thin-film systems. Going beyond the mechanical properties of coatings, Wang et al. [3] studied the viscoelastic behavior of memory chip 3D-stacked packaging, using finite element simulations to predict the effects of epoxy molding compound materials on electronic packaging reliability. In the realm of semiconductor packaging in harsh service conditions, Yoo and Kim [4] explored the galvanic corrosion behavior of Cu-based wire bonding, providing valuable insights into enhancing the longevity of electronic components exposed to corrosive environments. Coatings have also contributed much to the energy sector, where Drinčić et al. [5] analyzed the degradation processes of solar absorber coatings; for parabolic trough collectors, they suggested durable materials. Similarly, Yang et al. [6] studied NiO-doped Ga_2O_3 thin films, which have opened new prospects for enhancing performance in wide-bandgap semiconductors for energy-related devices.

Received: 8 February 2025
Accepted: 11 February 2025
Published: 12 February 2025

Citation: Long, X. Coatings for Advanced Devices. *Coatings* **2025**, *15*, 219. https://doi.org/10.3390/coatings15020219

Further extending the method of investigation and service scenarios of the coating materials and technologies, Zhao et al. [7] made a first-principles calculation of the mechanical properties of Ni_3Sn_4-based intermetallic compounds with cerium doping to highlight their potential for application in soldering in microelectronics. They also explored the microstructure and mechanical properties of TixNbMoTaW refractory high-entropy alloys, offering a novel approach to the development of robust coatings for extreme temperature environments, such as those encountered in high-performance aerospace and automotive applications [8]. Zou et al. [9] presented an innovative toughened bamboo-fiber-modified epoxy resin coating, which demonstrates superior interfacial compatibility, making it a promising candidate for bio-based, environmentally friendly polymer coatings. Eversole et al. [10] studied optical limiting in CdSe-based multiphase polymer nanocomposite films, further extending the role of coatings in optical applications by offering enhanced protection capabilities for optoelectronic devices under intense illumination.

Last but not least, this Special Issue is focused on coatings for enhancing performance, reliability, and sustainability in electronic devices in various domains. Papers in this Special Issue provide insights into different technologies and applications that will be shaping the functionality of advanced devices and systems in the future.

Conflicts of Interest: The author declares no conflicts of interest.

References

1. Long, X.; Li, J.; Shen, Z.; Su, Y. Dimensionless Analysis to Determine Elastoplastic Properties of Thin Films by Indentation. *Coatings* **2022**, *12*, 1768. [CrossRef]
2. Long, X.; Su, T.; Chen, Z.; Su, Y.; Siow, K.S. Tunable Coefficient of Thermal Expansion of Composite Materials for Thin-Film Coatings. *Coatings* **2022**, *12*, 836. [CrossRef]
3. Wang, X.; Cao, S.; Lu, G.; Yang, D. Viscoelastic Simulation of Stress and Warpage for Memory Chip 3D-Stacked Package. *Coatings* **2022**, *12*, 1976. [CrossRef]
4. Yoo, Y.R.; Kim, Y.S. Effects of Pd Alloying and Coating on the Galvanic Corrosion between Cu Wire and Bond Pads for a Semiconductor Packaging. *Coatings* **2024**, *14*, 544. [CrossRef]
5. Drinčić, A.; Noč, L.; Merzel, F.; Jerman, I. Future Parabolic Trough Collector Absorber Coating Development and Service Lifetime Estimation. *Coatings* **2024**, *14*, 1111. [CrossRef]
6. Yang, C.F.; Tsao, E.C.; Wang, Y.W.; Lin, H.P.; Meen, T.H.; Liao, S.H. Analyses of the Properties of the NiO-Doped Ga_2O_3 Wide-Bandgap Semiconductor Thin Films. *Coatings* **2024**, *14*, 1615. [CrossRef]
7. Zhao, R.; Cao, Y.; He, J.; Chen, J.; Liu, S.; Yang, Z.; Lin, J.; Chang, C. First-Principles Study on the Mechanical Properties of Ni_3Sn_4-Based Intermetallic Compounds with Ce Doping. *Coatings* **2025**, *15*, 59. [CrossRef]
8. Zhao, R.; Cao, Y.; He, J.; Chen, J.; Liu, S.; Yang, Z.; Lin, J.; Chang, C. Microstructure and Mechanical Properties of TixNbMoTaW Refractory High-Entropy Alloy for Bolt Coating Applications. *Coatings* **2025**, *15*, 120. [CrossRef]
9. Zou, B.; Huang, K.; Ma, J. Toughened Bamboo-Fiber-Modified Epoxy Resin: A Novel Polymer Coating for Superior Interfacial Compatibility. *Coatings* **2025**, *15*, 181. [CrossRef]
10. Eversole, L.M.; Adjorlolo, R.; Renaud, J.F.; Bhowmick, M. Optical Limiting from CdSe-Based Multiphase Polymer Nanocomposite Films. *Coatings* **2024**, *14*, 634. [CrossRef]

coatings

Article

Dimensionless Analysis to Determine Elastoplastic Properties of Thin Films by Indentation

Xu Long *, Jiao Li, Ziyi Shen and Yutai Su *

School of Mechanics, Civil Engineering and Architecture, Northwestern Polytechnical University, Xi'an 710072, China
* Correspondence: xulong@nwpu.edu.cn (X.L.); suyutai@nwpu.edu.cn (Y.S.)

Abstract: By assuming the elastoplastic properties of thin-film materials, a reverse analysis method is proposed by deriving a dimensionless function for the indentation process. The substrate effect is taken into account by assuming a perfect interface between thin-film and substrate materials. In order to obtain the applied load–penetration depth (*P-h*) curves, the indentation process is numerically modeled as an axisymmetric problem with a rigid-body Berkovich indenter on the semi-infinite substrate when performing finite element (FE) simulations. As a typical soft film/hard substrate problem, the elastic substrate is assumed and the power–law model is used to describe the constitutive properties of thin-film materials. Varying elastic modulus (10–50 GPa), yield strength (60–300 MPa), and hardening exponent (0.1–0.5) characterize different elastoplastic mechanical properties of thin-film materials with film thickness of 10–30 μm. Owing to the good trending *P-h* curves with the maximum indentation depth up to the 2/3 film thickness for different elastoplastic thin-film materials, a dimensionless function is derived and validated based on the predictions by reliable FE simulations. The proposed dimensionless function elegantly elucidates the essential relationship between the elastoplastic mechanical properties of the thin-film material and indentation responses (e.g., loading and unloading variables). The elastoplastic constitutive curves predicted by the proposed reverse method are confirmed to be in good agreement with the stress-strain curves of materials by FE simulations with the randomly selected elastoplastic mechanical properties and film thicknesses. This study provides a theoretical guidance to understand the explicit relationship between elastoplastic mechanical properties of the thin-film material and indentation responses.

Keywords: indentation; thin-film; reverse analysis; dimensionless function; finite element simulation

Citation: Long, X.; Li, J.; Shen, Z.; Su, Y. Dimensionless Analysis to Determine Elastoplastic Properties of Thin Films by Indentation. *Coatings* **2022**, *12*, 1768. https://doi.org/10.3390/coatings12111768

Academic Editor: Angela De Bonis

Received: 23 October 2022
Accepted: 15 November 2022
Published: 18 November 2022

1. Introduction

Thin film/substrate systems are of significance in numerous critical engineering applications such as micro-electronics, optoelectronics, display panels, and many other devices, as illustrated in Figure 1. Processing techniques, for instance, sputtering, vapor deposition, ion implantation, and laser glazing are employed to fabricate thin film/substrate systems [1–3]. The reliability of microelectronic devices relies primarily on how reliably the thin films adhere, or "stick" to each other and substrates [4], thus the adhesion measurement of thin film has been of great interest in recent years. Carbon nanotubes and graphene are successfully applied to energy conversion systems, including solar cells and fuel cells, due to their excellent structures and properties in many vital areas [5]. A new domain of technical physics, integrated optics, is stimulated by the technical problems related to the widespread application of laser technology, both for actual aims (communication, information processing, ranging, etc.,) and for physical research [6]. In addition, there are various applications, from computer hardware and sensors to thin films and coatings, where important parts are manufactured in small sizes and low thicknesses [7,8]. Depth-sensing nanoindentation measurement techniques are frequently utilized to determine the mechanical properties of surface layers and ultrathin coatings of bulk materials [9]. In fact,

d owing to the impressive

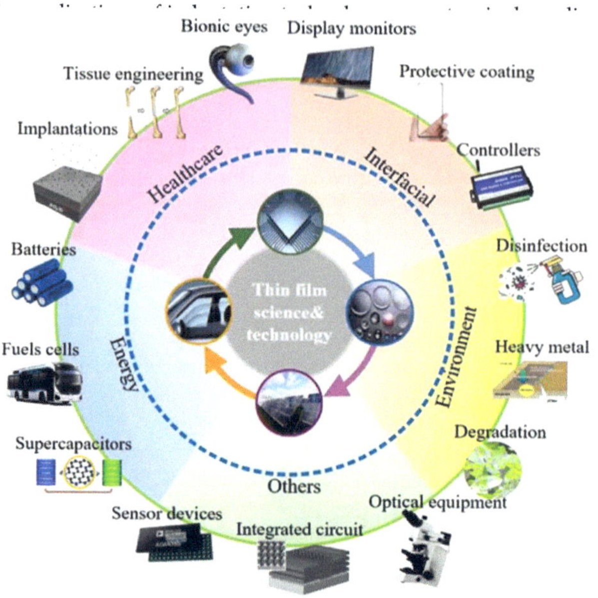

Figure 1. Schematic diagram of thin-film applications in various areas.

The method proposed by Oliver and Pharr in 1992 to measure hardness and elastic modulus by indention techniques, has extensively been applied and used for characterization of small-scale mechanical behavior [10]. In recent decades, because of its good performance in characterizing the mechanical properties of each material at low scales, the nanoindentation theory technique has been applied to measure the mechanical properties of micro-electronics and nano-scale component materials as well for local mechanical properties of materials [11]. For anisotropic materials, the question is what modulus is measured by the indentation technique due to load and displacement [12]. On the microscopic aspect, Bouzakis et al. [13] investigated the effect of surface roughness of the substrate material due to different size shapes in nanoindentation experiments based on FE simulation. Rauchs [14] performed nanoindentation experiments to determine the material parameters of an elastic-viscoplastic material model with nonlinear isotropy and kinematic hardening by minimizing the least-squares difference between experimental data from indentation tests and FE predictions. Shelef and Bar-On [15] related the indentation modulus magnitude and loss parameters to the modulus of the original film to derive the film-substrate energy storage and energy dissipation capabilities, and outlines a method to back-calculate the elements by simple linear scaling. Kampouris et al. [16] presented a new approach to interpreting Vickers microindentation data, using a gradient elastic framework fitting formulation, and showed experimentally the potential of the formulation to model such microdeformation issues. With regard to improving the accuracy of material parameters measured by nanoindentation, Chollacoop et al. [17] proposed an empirical method to improve the accuracy of material parameter measurement by enriching the indentation method with utilizing double sharp indenters with different tip-top angle. Treier et al. [18] combined the genetic algorithm with the requested parameters and applied to the identification of material parameters. Kim et al. [19] proposed a size effect model by investigating the effect of roughness in nanoindentation experiments. Based on the first-order asymptotic solution, a simple analytical approximation is suggested for the indentation scaling factor that takes into account the elastic layer's finite thickness as well as the effect of the elastic substrate [20]. By conducting indentation experiments, Long et al. [21–24] investigated the residual stresses in Sn-3.0Ag-0.5Cu lead-free solder after annealing treatment, and discussed the effects of different annealing temperatures and durations on the applied load–penetration depth (*P-h*) curves. Long et al. [25] established a quantitative relationship between indentation strain rate and uniaxial tensile strain rate by introducing the concept of rate factor, which is based on a set of polynomial functions connecting the instantaneous and residual plastic zone radii and helps to break the uniqueness of determining the unique elastic-plastic properties of a material by indentation. Researchers focus on the effect of surface stress by pre-stressing the materials to be indented by a Berkovich indenter [26]. In order to improve the accuracy of the reverse calculation, a great deal of work has been done using different shapes of material properties of the indenter [27–29], proposing advanced algorithms for FE simulations [30,31]. By using dimensionless analysis, FE results are first combined with dimensionless functions, and then a series of nonlinear fitting equations are proposed to determine the dimensionless equations for the mechanical properties of elastic-plastic solids with work hardening [32]. It is noticed that these studies extended the application of indentation to characterize mechanical properties such as surface stresses, residual stresses, and plastic properties of elastoplastic substrates. However, existing studies were limited to the cases with only the substrate model. This means that the methods and conclusions from existing studies cannot shed light on the cases for film/substrate model. This underlines the urgent necessity to incorporate thin-film materials in the FE model and consider the influence of the presence of thin films on the whole process of indentation.

The purpose of this paper is to develop a method involving a combination of indentation techniques and FE simulations to describe the mechanical behavior of thin films. The reliability of the modeling is investigated mainly through simulation to verify the experimental phenomena. The film is further assigned on the basis of a pure substrate model in order to find out the influence of film properties on the indentation process. Finally, a reverse analysis based on *P-h* curves as indentation response is presented in this paper to acquire the mechanical properties of the substrate material. However, the theory of the indentation technique is not well established especially for thin film materials. Hence, a dimensionless algorithm based on the *P-h* curve as the indentation response is proposed to obtain the mechanical properties of the thin film material.

2. Indentation Simulation by FE Method

2.1. Validation with No-Film Model

As thin metal films are widely used in microelectronic devices, optics, and other fields, the mechanical properties of the films are also required in practice. Hence, it is especially important to experimentally and numerically study the mechanical properties of thin films. The indentation method can better address these problems and obtain the measurement results with sufficient accuracy. However, the indentation method inevitably brings some errors in the process of testing, such as local wear of the indenter, microscopic defects on the surface of the specimen, and small changes in the experimental environment. These minor error sources can be neglected in macro-scale studies. But these error sources can greatly reduce the accuracy of measurement results in the nano-scale. With the help of FE analysis, the application of indentation characterization of various mechanical theoretical models can be significantly improved, by improving the mechanical properties of materials in terms of measurement accuracy. Therefore, the information provided by FE simulations combined with indentation tests can be more comprehensive. By performing FE simulations, the predictions based on the three-dimensional Berkovich model are in general accordance with those by using the two-dimensional equivalent model as found by Zhuk et al. [32]. Essentially, the standard Berkovich indenter enables the consistency in terms of projected area by equivalent sharp indenters (e.g., conical, triangular pyramidal and square-based pyramidal indenters) for the indentations [33,34].

Figure 2 shows the schematic diagram of the FE model for indention simulation with a film/substrate structure by using a Berkovich indenter. The film is attached on top of the substrate by assuming a perfect interfacial adhesion. This means that no delamination failure is taken into account in the present FE model. Regarding the loading condition, displacement control is applied on the indenter by means of a reference point. For the Berkovich indenter, the FE model can be simplified as an axisymmetric model with the indentation half-angle of 70.3°. The substrate is regarded as a semi-infinite space body with a height of 3000 μm, and the film thicknesses are selected as 10 μm, 20 μm, and 30 μm for the parameter identification of subsequently proposed dimensionless functions, and the maximum penetration depth is selected as 2/3 of the film thickness. The mesh discretization of the FE model is confirmed to be small enough to simulate such a semi-infinite space problem after performing mesh sensitivity study.

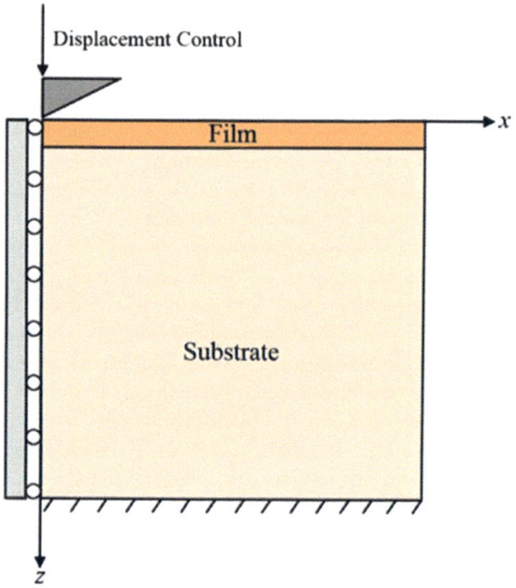

Figure 2. Schematic diagram of the indentation model with a film/substrate structure under a Berkovich indenter.

For the axisymmetric boundary condition, appropriate constraints are applied to the left and lower sides in the semi-infinite substrate to simulate the real experiments. As the stress distribution in the substrate induced by the indenter penetration is localized, the left side of the semi-infinite substrate is free. The vertical displacement is applied at the reference point of the indenter to penetrate the indenter into the film/substrate structure. The applied load P on the indenter and the penetration depth h can be recorded throughout the indentation process, so that the P-h curves under different working conditions are utilized to characterize the mechanical response of the film/substrate structure. In the elastoplastic deformation process, the substrate is considered to be elastic as the induced stress is not significant, while the film material is identified by a piecewise elastoplastic model with the plastic stage described by a power-law model which is usually expressed as

$$\sigma = \begin{cases} E\varepsilon & \varepsilon \leq \varepsilon_y \\ R\varepsilon^n & \varepsilon > \varepsilon_y \end{cases}, \tag{1}$$

where E is the Young's modulus, R is the strength coefficient, n is the hardening exponent, and the yield strain ε_y is the corresponding to the yield strength which can be expressed as

$$\sigma_y = E\varepsilon_y = R{\varepsilon_y}^n, \tag{2}$$

The total strain can be represented by the sum of the elastic strain ε_e and the plastic strain ε_p as

$$\varepsilon = \varepsilon_e + \varepsilon_p, \tag{3}$$

Combining the above equations, the stress σ can be written as

$$\sigma = \frac{\sigma_y}{\varepsilon_y{}^n}(\varepsilon_e + \varepsilon_p). \tag{4}$$

For validating the established FE models for simulating indentation response, a series of comparisons is made for the FE predictions with the reported experimental results in the literature. First, considering a simpler indentation condition without the composition of film and substrate in the FE model, the predicted *P-h* curves are compared with the experimental results [34]. Table 1 lists the mechanical properties for both elastic Berkovich indenter and elastoplastic substrate. A wide range of film materials is elastoplastic materials with the film thickness of 10–30 μm and the mechanical properties for Young's modulus of 10–50 GPa, yield strength of 60–300 MPa, hardening exponent of 0.1–0.5, and Poisson's ratio of 0.07. As shown in Figure 3, the solid line indicates the FE simulations and the spheres represent the experimental measurements [34], both of which are in good agreement, confirming the accuracy of the FE models when predicting the indentation response.

Table 1. Mechanical properties for indenter and substrate.

-	*E* (GPa)	v	n	σ_{fy} (Mpa)
Berkovich indenter	1060	0.07	-	-
Substrate	215	0.28	0.13	330

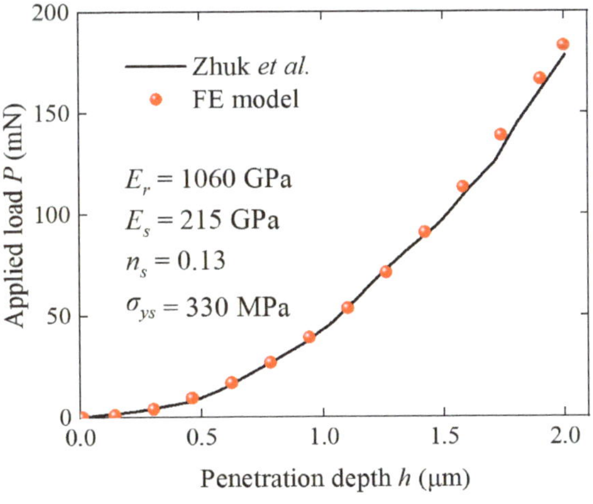

Figure 3. Response comparison of the established FE model and published result [34].

In addition to the validation of loading stage, the full process for both loading and unloading is adopted to validate the indentation prediction accuracy of the established FE model with the published results [35], as shown in Figure 4. In this validation, the maximum value of penetration depth is 200 nm. For substrate, the Young's modulus is 65 GPa and the yield strength is 300 MPa. For indenter, the Young's modulus is 1141 GPa and the Poisson's ratio is 0.07. As shown in Figure 4, compared with the validation case with 2 μm, the penetration depth of 200 nm is much smaller. This means the established FE models are nano and micro-scales capable of predicting the indentation behavior for both nano and micro-scales.

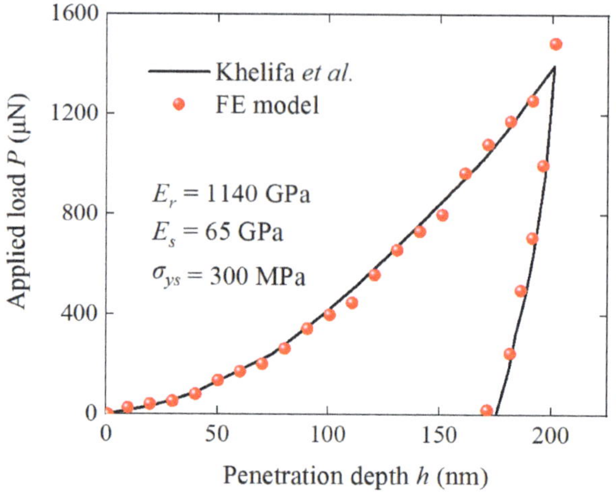

Figure 4. Response comparison of the established FE model and published result [35].

2.2. Validation with Film/Substrate Model

After validating the FE model for the indentation of homogenous substrate materials, the experimental result of composite materials with both film and substrate [36] are also taken into account for validating the prediction accuracy of the established thin film/substrate model. The indenter material is diamond with Young's modulus of 1141 GPa, and Poisson's ratio of 0.07. The film material is TiN with the thickness of 2 μm, Young's modulus of 427 GPa, Poisson's ratio of 0.25, and yield strength of 13,500 MPa. For the substrate, the Young's modulus is 218 GPa, Poisson's ratio is 0.3, yield strength is 1800 MPa. With the good agreement of *P-h* curves as shown in Figure 5, the established FE model is confirmed to be satisfactory to predict the indentation behavior of thin film/substrate structure. Moreover, after further examining the stress distribution in film/substrate as shown in the insert of Figure 6, severe plastic deformation is induced in the thin film but elastic deformation is dominant for the substrate materials. So, it is reasonable to assume elastic substrate material and focus on the elastoplastic behavior of thin-film materials in the subsequent FE simulations with film/substrate of the present study.

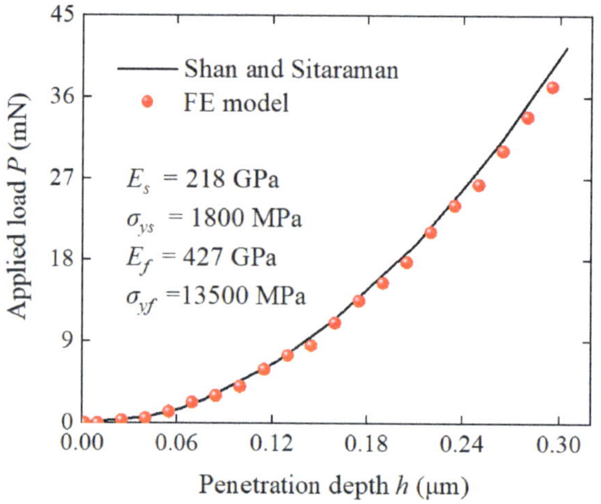

Figure 5. Response comparison of the established FE model and the published result [36].

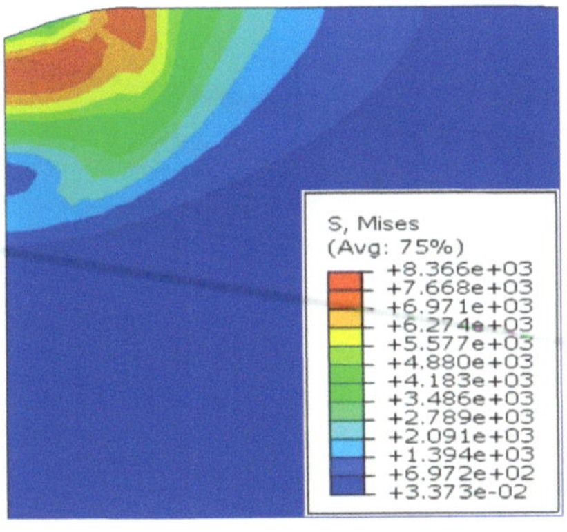

Figure 6. Stress deformation diagram for indentation depth of 0.3 μm.

Subsequently, the indentation on hard film/soft substrate and soft film/hard substrate material systems was comprehensively investigated. The soft and hard materials are defined as isotropic elastic-perfectly plastic materials. Young's modulus and yield strength of the soft material are taken as 100 GPa and 1000 MPa, respectively. The corresponding values for the hard material are 200 GPa and 20,000 MPa. The depth of substrate is 200 μm, while the thickness of thin films is 2 μm. The distributions of equivalent plastic strain in both cases are compared in Figures 7 and 8. It was found that the deformation and the corresponding value PEEQ under different working conditions are very close to each other.

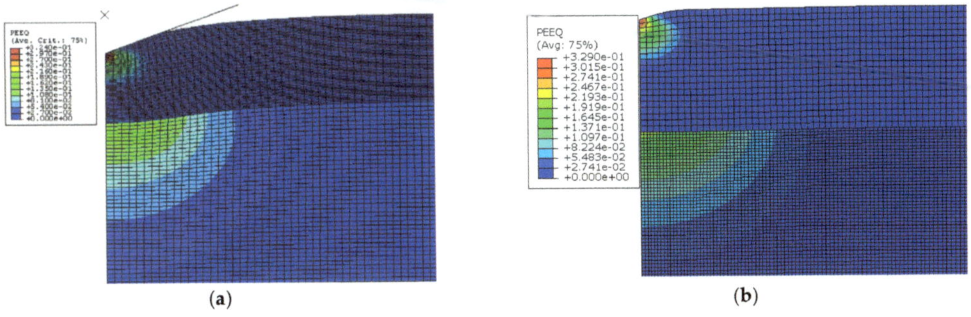

Figure 7. PEEQ comparison in the hard film/soft substrate model at h_{max} = 900 nm: (**a**) reported result by Pelegri et al. [37] (maximum value is 0.3240); (**b**) FE result (maximum value is 0.3290).

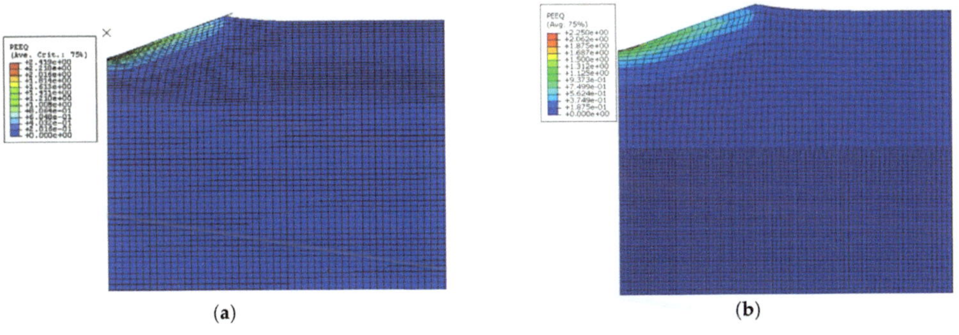

Figure 8. PEEQ comparison in soft film/hard substrate model at h_{max} = 900 nm: (**a**) reported result by Pelegri et al. [37] (maximum value is 2.419); (**b**) FE result (maximum value is 2.250).

3. Dimensionless Analysis

Indentation tests are versatile in terms of controlling the applied displacement, load, and time sequence, which can be illustrated in Figure 9. An indentation process includes usually at least a sequence of loading and unloading. In the local region underneath the indenter, elastic and plastic deformations occur when the indenter is penetrated into the substrate material until the maximum penetration depth h_{max}. The indentation develops with a shape corresponding to the indenter shape. Since the plastic part does not fully recover after the complete unloading of the load on the indenter, only the part of the elastic deformation can be recovered until the residual penetration depth h_r. Elastic properties can be extracted from the elastic part of the unloading curve. This will facilitate the use of the contact stiffness S as the initial slope of the unloading curve to quantify the elastic properties such as the Young's modulus. Considering the composition of film and substrate of interest in the present study, the mechanical properties of the films can be further elucidated by proposing the dimensional analysis based on extensive FE simulations.

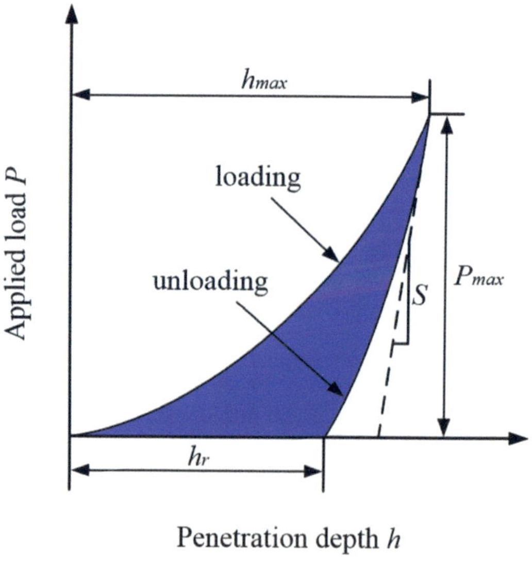

Figure 9. Typical *P-h* curve under the thin film/substrate model.

The continuous measurement of the applied load during the loading stage can be utilized to measure the hardness. As the average contact pressure of the material under the indenter [37], hardness exhibits the material ability to resist local deformation, especially plastic deformation, indentation, or scratching. Conventionally, hardness is defined by

$$H = \frac{P}{A_c}. \tag{5}$$

where P is the applied load that is obtained directly from the *P-h* curve, and A_c is the projected contact area of the indenter tip in the substrate material which can be determined for a Berkovich indenter as

$$A_c = 24.5h^2. \tag{6}$$

For a conical indenter in this axisymmetric problem, the contact stiffness S can be used to calculate the Young's modulus of the material and denoted as

$$\mathrm{S} = \frac{\mathrm{d}P}{\mathrm{d}h} = \frac{2}{\sqrt{\pi}} E_r \sqrt{A_c}. \tag{7}$$

where $\mathrm{d}P/\mathrm{d}h$ is the initial slope of the unloading curve, and E_r is the reduced Young's modulus of the material which can be defined as

$$\frac{1}{E_r} = \frac{1 - {v_s}^2}{E_s} + \frac{1 - {v_i}^2}{E_i}. \quad (8)$$

The reduced Young's modulus E_r is evaluated from the indentation measurements according to the following equation

$$E_r = \frac{\sqrt{\pi}}{2} \frac{(\mathrm{d}P/\mathrm{d}h)_{\text{unload}}}{\beta\sqrt{A_c}}. \quad (9)$$

where $(\mathrm{d}P/\mathrm{d}h)_{\text{unload}}$ is the scope of the unloading curve evaluated at the position of maximum load. β is a dimensionless parameter related to the geometry of the indenter. For the case of tapered indenter β = 1 [38].

However, it should be noted that the above equations cannot be applied directly for the mechanical characterizations of thin films which are supported by their substrates, as the substrate effect cannot be neglected. Usually, in order to avoid the influence of the substrate for measuring reliable mechanical properties, the penetration depth should be less than 10% of the film thickness. On the other hand, if the penetration depth is too small by satisfying this limitation, the size effect should be concerned for such shallow indentations. Therefore, the above equations can still be used but with a compromised penetration depth for thin-film materials. This works is motivated to propose a method to determine the mechanical properties of thin-film materials by considering the substrateg effect.

The applied load P generated by the entire indentation process is extremely dependent on the mechanical parameters between the substrate and the film. As revealed by the FE predictions, similar trend is expected to be captured by the proposed dimensionless function in a similar way of the works done by Long et al. [39,40]. During the elastic-plastic indentation deformation, the applied load P can be given as

$$P = P\left(E_s, E_f, n_f, \sigma_{fy}, h, t\right), \quad (10)$$

where E_s is the Young's modulus of the base material, E_f and n_f are the Young's modulus and hardening exponent of the Berkowitz indenter, σ_{fy} is the yield strength of the film, h is the penetration depth of the indenter, and t is the thickness of the film. Since the indenter parameters are constant, the material and geometric parameter related with the indenter is not considered.

As identified in Equation (10), seven variables with three independent dimensions (that is, length, mass, and time) are utilized. By proposing four dimensional terms, the dimensionless functions are rewritten as

$$\frac{P}{E_f \cdot h^2} = \prod\left(\frac{n_f \cdot E_f}{E_s}, \frac{\sigma_{fy}}{E_f}, \frac{h}{t}\right). \quad (11)$$

By identifying the values of the dimensionless parameters, the dimensionless formulae are derived in an elegant way to better reproduce the FE predictions. In order to increase the reliability of the parameter fitting for the proposed dimensional analysis, random work conditions with statistical distributions are considered when preparing the database of FE simulations. For the selection of working conditions in this study, in addition to varying the values with certain increments to cover the whole range of mechanical parameters, some more working cases are also determined randomly so as to enrich the generalization of the indentation database. For instance, the histogram of the determined thickness t is shown in Figure 10. The other parameters such as Young's modulus, yield strength, and hardening exponent are also randomly determined in their individual ranges. The indentation database with both incremental and random conditions can effectively enhance the parameter fitting of the proposed dimensionless function.

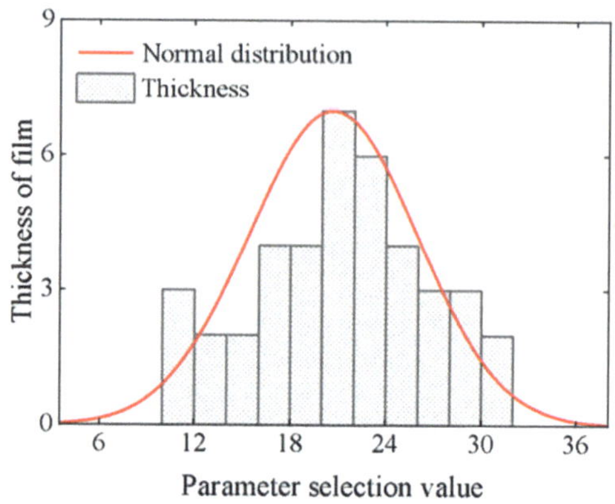

Figure 10. Distribution histogram of the thickness randomly determined in the range of interest.

4. Results and Discussions

Young's modulus of the substrate is much larger than that of the film, which is the standard case for the soft film/hard substrate of interest in this study. For the parameter selection of FE simulations, the reasonable ranges of mechanical properties for the thin-film materials are E_f = 10–50 GPa, σ_{yf} = 60–300 MPa, n = 0.1–0.5, and t = 10–30 μm, while the substrate material remains elastic. The variation of these mechanical parameters covers most of the thin-film materials in the industries. The film constitutive behavior is described according to Equation (1). After performing extensive FE simulations, the effects of mechanical properties and film thickness on the *P*-*h* curves are further investigated.

As illustrated in Figure 11, it can be seen that the effects of these four parameters on the curves are consistent. The applied load exponentially increases with the increase of mechanical properties. As the penetration depth increases, greater difference is found for the applied load due to the increase of these parameters gradually. Moreover, the applied load decreases as the film thickness increases.

Apparently, the influences of these factors of material parameters and film thickness can be well described by exponential fitting functions, which significantly guides the explicit form of the proposed dimensional function.

In this work, the commercial analysis software Origin 2018 64 Bit is used to fit the dimensionless formula. By focusing on the effect of material parameters and film thickness on the corresponding *P*-*h* curve during the indentation, the indentation problem is symmetrically investigated by performing FE analysis. Based on the indentation method proposed by Oliver and Pharr [10] and the elastic-fully plastic material model, mechanical parameters have been assumed. To perform the reverse analysis of constitutive parameters of thin-film materials based on the indentation responses, a parameterized dimensionless function is achieved based on the obtained FE results as

$$\frac{P}{E_f \cdot h^2} = a \cdot \frac{\sigma_{fy}}{E_f} \cdot \left(1 + b \cdot \frac{h}{t}\right) \cdot \exp\left(c \cdot \frac{n_f \cdot E_f}{E_s}\right) + d \cdot \frac{h}{t}. \quad (12)$$

where $a = 0.04797$, $b = 1.19646$, $c = 2.82563$, and $d = 3.87933 \times 10^{-4}$. Furthermore, the fitting accuracy is measured by the coefficient of determination R^2 of 0.90 for the FE predictions in the database of various working conditions. This proves that the dimensionless function agrees well with the FE simulations. It should be noted that the proposed dimensionless function can be utilized to guide the parameters tuning when performing indentation tests or simulations on film/substrate structures.

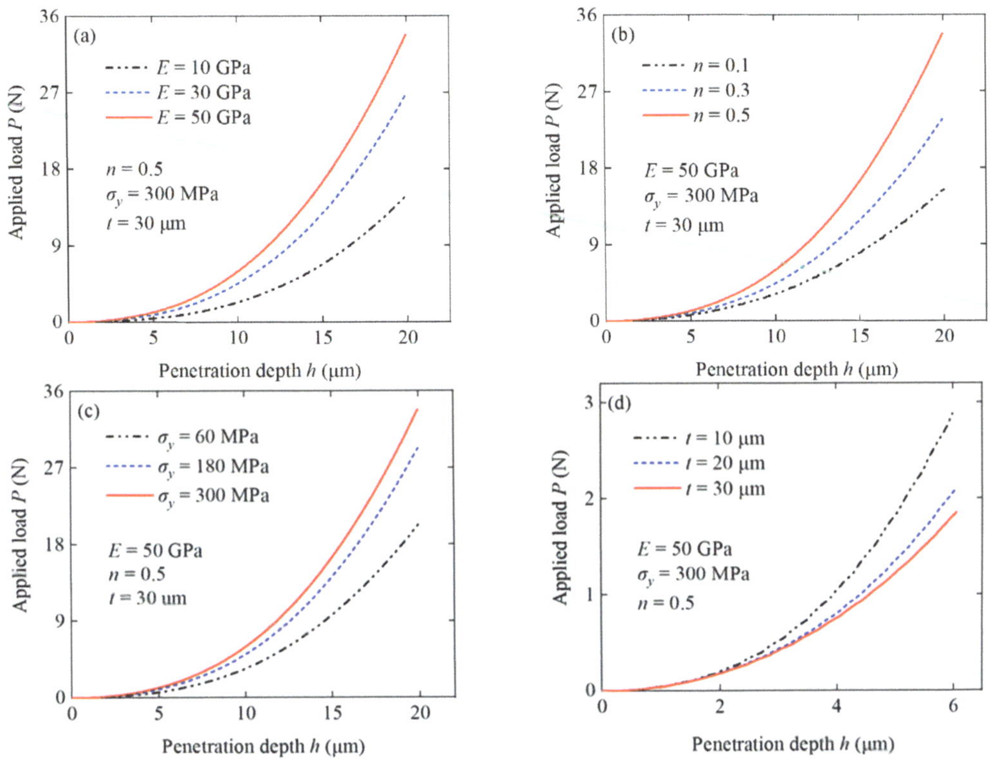

Figure 11. Effect of material parameters and film thickness on the *P*-*h* responses: (**a**) Young's modulus; (**b**) hardening exponent; (**c**) yield strength; (**d**) film thickness.

In order to further examine the reliability of the dimensionless formulation, three working conditions are randomly determined as listed in Table 2 (Case1: E = 41,905.8 MPa, n = 0.48, σ_{fy} = 252.5 MPa, t = 24 μm; Case2: E = 20,251.6 MPa, n = 0.40, σ_{fy} = 225.7 MPa, t = 20 μm; Case3: E = 31,448.7 MPa, n = 0.19, σ_{fy}=112.4 MPa, t = 13 μm). It should be noted that all cases are out of the database for the parameter identification. Thus, the comparisons between the predictions by the proposed dimensional method and the established FE model can be utilized to confirm the prediction accuracy of the proposed dimensional method.

Table 2. Three new randomly determined working conditions.

-	E (MPa)	n	σ_{fy} (Mpa)	t (μm)
Case1	41,905.8	0.48	252.5	24
Case2	20,251.6	0.40	225.7	20
Case3	31,448.7	0.19	112.4	13

The solid line denotes the FE simulation results, and the scattered spheres corresponds to the predicted results at different penetration depths. As shown in Figure 12, the final data show that the dimensionless formulation is highly predictive for FE simulations, testifying to the credibility of Equation (12).

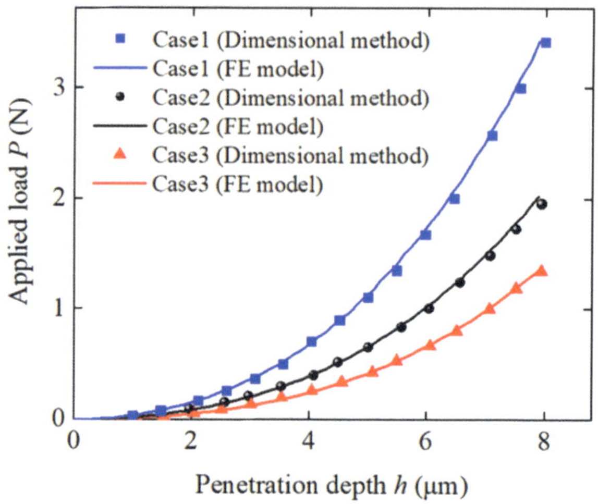

Figure 12. Prediction comparisons by the proposed dimensional method and the established FE model for three new randomly determined conditions.

5. Conclusions

A dimensionless analysis is proposed to estimate the elastoplastic properties of thin-film materials by the indentation method. After the simulation validation of the published results, an FE method is used to parametrically investigate the extensive mechanical parameters of the thin-film material. The prediction of the FE results shows a consistent loading trend for various mechanical properties and film thicknesses, which makes it possible to explore the relationship between the individual variables. With the parameters to describe the indentation problem for dimensional analysis, a dimensionless function related to the applied load is proposed. Using the proposed dimensionless function, the penetration depth within 6 μm is fitted with FE predictions up to 0.90. This means that unlike simulations, the dimensionless function provides a reliable method of analysis. With the film material and its mechanical properties established, the thickness of the film required to withstand external loads may be a practical application technique that can be developed. The applied loads were estimated for the films with different material properties and the obtained results fit well with the results. FE simulations can verify the accuracy of experiments, shorten the experimental cycle, and reduce material losses, and the dimensionless method has good application prospects and provides theoretical guidance for establishing parameter relationships by numerical methods.

Author Contributions: X.L.: conceptualization, methodology, funding acquisition, supervision, writing—review and editing; J.L.: investigation, methodology, writing—original draft, visualization, software; Z.S.: methodology, writing—original draft, visualization; Y.S.: methodology, software, writing—review and editing. All authors have read and agreed to the published version of the manuscript.

Funding: This work was supported by the National Natural Science Foundation of China (No. 52175148), and the Natural Science Foundation of Shaanxi Province (No. 2021KW-25).

Institutional Review Board Statement: Not applicable.

Informed Consent Statement: Not applicable.

Data Availability Statement: The data presented in this study are available on request from the corresponding author. The data are not publicly available due to privacy.

Conflicts of Interest: The authors declare no conflict of interest.

References

1. Lin, K.L.; Chang, W.K.; Chang, T.C.; Lee, C.H.; Lin, C.H. Recycling thin film transistor liquid crystal display (TFT-LCD) waste glass produced as glass-ceramics. *J. Clean Prod.* **2009**, *17*, 1499–1503. [CrossRef]
2. Nagamura, T.; Tanaka, K. Physical properties of polymer thin Films for device application. *IETE Tech. Rev.* **2009**, *109*, 1–2.
3. Gizatulin, A.R.; Sultanov, A.K. Application of whispering gallery modes (WGM) in optical communications. *Opt. Commun.* **2017**, *10774*, 545–555.
4. Cordill, M.J.; Bahr, D.F.; Moody, N.R.; Gerberich, W.W. Recent developments in thin film adhesion measurement by laser-induced stress waves. *J. Exp. Med.* **2007**, *22*, 249–257.
5. Ouyang, J. Applications of carbon nanotubes and graphene for third-generation solar cells and fuel cells. *Nano Mater. Sci.* **2019**, *1*, 77–90. [CrossRef]
6. Prokhorov, A.M.; Smolenskiĭ, G.A.; Ageev, A.N. Optical phenomena in thin-film magnetic waveguides and their technical application. *Sov. Phys. Uspekhi* **1984**, *27*, 339. [CrossRef]
7. Islam, M.M.; Shakil, S.I.; Shaheen, N.M.; Bayati, P.; Haghshenas, M. An overview of microscale indentation fatigue: Composites, thin films, coatings, and ceramics. *Micron* **2021**, *148*, 103110. [CrossRef]
8. Zhou, Y.; Lin, X.G.; Wang, Y.; Liu, G.Q.; Zhu, X.Y.; Huang, Y.K.; Guo, Y.C.; Gao, C.; Zhou, M. Study on gas sensing of reduced graphene oxide/ZnO thin film at room temperature. *Sens. Actuators B Chem.* **2017**, *240*, 870–880. [CrossRef]
9. Bhushan, B.; Li, X.D. Nanomechanical characterisation of solid surfaces and thin films. *Int. Mater. Rev.* **2003**, *48*, 125–164. [CrossRef]
10. Oliver, W.C.; Pharr, G.M. An improved technique for determining hardness and elastic modulus using load and displacement sensing indentation experiments. *J. Mater. Res.* **1992**, *7*, 1564–1583. [CrossRef]
11. He, D.; Zhu, J.C.; Lai, Z.H.; Liu, Y.; Yang, X.W.; Nong, Z.S. Residual elastic stress-strain field and geometrically necessary dislocation density distribution around nano-indentation in TA15 titanium alloy. *Trans. Nonferrous Met. Soc. China* **2013**, *23*, 7–13. [CrossRef]
12. Oliver, W.C.; Pharr, G.M. Measurement of hardness and elastic modulus by instrumented indentation: Advances in understanding and refinements to methodology. *J. Mater. Res.* **2004**, *19*, 3–20. [CrossRef]
13. Bouzakis, K.D.; Michailidis, N.; Hadjiyiannis, S.; Skordaris, G.; Erkens, G. The effect of specimen roughness and indenter tip geometry on the determination accuracy of thin hard coatings stress-strain laws by nanoindentation. *Mater. Charact.* **2002**, *49*, 149–156. [CrossRef]
14. Rauchs, G.; Bardon, J. Identification of elasto-viscoplastic material parameters by indentation testing and combined finite element modelling and numerical optimization. *Finite Elem. Anal. Des.* **2011**, *47*, 653–667. [CrossRef]
15. Shelef, Y.; Bar-On, B. Size effects on the dynamic indentation modulus of films. *Mech. Mater.* **2022**, *164*, 104118. [CrossRef]
16. Kampouris, A.K.; Lappas, K.I.; Konstantinidis, A.A.; Aifantis, E.C. A new method for interpreting Vickers indentation measurements. *Mater. Today* **2022**, *67*, 964–970. [CrossRef]
17. Chollacoop, N.; Dao, M.; Suresh, S. Depth-sensing instrumented indentation with dual sharp indenters. *Acta Mater.* **2003**, *51*, 3713–3729. [CrossRef]
18. Treier, K.; Lester, P.; Hubbuch, J. Application of genetic algorithms and response surface analysis for the optimization of batch chromatographic systems. *Biochem. Eng. J.* **2012**, *63*, 66–75. [CrossRef]
19. Kim, J.Y.; Kang, S.K.; Lee, J.J.; Jang, J.I.; Lee, Y.H.; Kwon, D. Influence of surface-roughness on indentation size effect. *Acta Mater.* **2007**, *55*, 3555–3562. [CrossRef]
20. Argatov, I.I.; Sabina, F.J. Small-scale indentation of an elastic coated half-space: Influence of Poisson's ratios on the substrate effect. *Int. J. Eng. Sci.* **2014**, *81*, 33–40. [CrossRef]
21. Long, X.; Zhang, X.D.; Tang, W.B.; Wang, S.B.; Feng, Y.H.; Chang, C. Micromachines calibration of a constitutive model from tension and nanoindentation for lead-free solder. *Micromachines* **2019**, *9*, 608. [CrossRef] [PubMed]
22. Long, X.; Wang, S.B.; Feng, Y.H.; Yao, Y.; Keer, L.M. Annealing effect on residual stress of Sn-3.0Ag-0.5Cu solder measured by nanoindentation and constitutive experiments. *Mat. Sci. Eng.* **2017**, *696*, 90–95. [CrossRef]
23. Long, X.; Wang, S.B.; He, X.; Yao, Y. Annealing optimization for tin-lead eutectic solder by constitutive experiment and simulation. *J. Mater. Res.* **2017**, *32*, 3089–3099. [CrossRef]
24. Long, X.; Feng, Y.H.; Yao, Y. Cooling and annealing effect on indentation response of lead-free solder. *Int. J. Appl. Mech.* **2017**, *9*, 1750057. [CrossRef]
25. Long, X.; Shen, Z.Y.; Jia, Q.P.; Li, J.; Dong, R.-P.; Su, Y.-T.; Yang, X.; Zhou, K. Determine the unique constitutive properties of elastoplastic materials from their plastic zone evolution under nanoindentation. *Mech. Mater.* **2022**, *175*, 104485. [CrossRef]
26. Long, X.; Shen, Z.Y.; Lu, C.H.; Jia, Q.P.; Guan, C.; Chen, C.T.; Wang, H.D.; Li, Y. Reverse analysis of surface strain in elasto-plastic materials by nanoindentation. *Int. J. Appl. Mech.* **2021**, *13*, 2150106. [CrossRef]
27. Chen, H.; Cai, L.X. Theoretical model for predicting uniaxial stress-strain relation by dual conical indentation based on equivalent energy principle. *Acta Mater.* **2016**, *121*, 181–189. [CrossRef]
28. Hosseinzadeh, A.R.; Mahmoudi, A.H. Determination of mechanical properties using sharp macro-indentation method and genetic algorithm. *Mech. Mater.* **2017**, *114*, 57–68. [CrossRef]
29. Hyun, H.C.; Kim, M.; Lee, J.H.; Lee, H. A dual conical indentation technique based on FEA solutions for property evaluation. *Mech. Mater.* **2011**, *43*, 313–331. [CrossRef]

30. Campbell, J.E.; Thompson, R.P.; Dean, J.; Clyne, T.W. Experimental and computational issues for automated extraction of plasticity parameters from spherical indentation. *Mech. Mater.* **2018**, *124*, 118–131. [CrossRef]
31. Li, Y.G.; Kanouté, P.; Francois, M.; Chen, D.; Wang, H.W. Inverse identification of constitutive parameters with instrumented indentation test considering the normalized loading and unloading *P-h* curves. *Int. J. Solids Struct.* **2019**, *156*, 163–178. [CrossRef]
32. Cheng, Y.T.; Cheng, C.M. Scaling approach to conical indentation in elastic-plastic solids with work hardening. *J. Appl. Phys.* **1998**, *84*, 1284–1291. [CrossRef]
33. Dao, M.; Chollacoop, N.; Vliet, K.; Venkatesh, T.A.; Suresh, S.J.A.M. Computational modeling of the forward and reverse problems in instrumented sharp indentation. *Acta Mater.* **2001**, *49*, 3899–3918. [CrossRef]
34. Zhuk, D.I.; Isaenkova, M.G.; Perlovich, Y.A.; Krymskaya, O.A. Finite element simulation of microindentation. *Russ. Metally* **2017**, *2017*, 390–396. [CrossRef]
35. Khelifa, M.; Fierro, V.; Celzard, A. Finite element simulation of nanoindentation tests using a macroscopic computational model. *J. Mech. Sci. Technol.* **2014**, *28*, 3209–3217. [CrossRef]
36. Shan, Z.; Sitaraman, S.K. Elastic-plastic characterization of thin films using nanoindentation technique. *Thin Solid Films* **2003**, *437*, 176–181. [CrossRef]
37. Pelegri, A.A.; Huang, X. Nanoindentation on soft film/hard substrate and hard film/soft substrate material systems with finite element analysis. *Compos. Sci. Technol.* **2008**, *68*, 147–155. [CrossRef]
38. Lichinchi, M.; Lenardi, C.; Haupt, J.; Vitali, R. Simulation of Berkovich nanoindentation experiments on thin films using finite element method. *Thin Solid Films* **1998**, *333*, 240–248. [CrossRef]
39. Long, X.; Jia, Q.P.; Li, Z.; Wen, S.X. Reverse analysis of constitutive properties of sintered silver particles from nanoindentations. *Int. J. Solids Struct.* **2020**, *191*, 351–362. [CrossRef]
40. Long, X.; Hu, B.; Feng, Y.H.; Chang, C.; Li, M.Y. Correlation of microstructure and constitutive behaviour of sintered silver particles via nanoindentation. *Int. J. Mech. Sci.* **2019**, *161*, 105020. [CrossRef]

coatings

MDPI

Article

Tunable Coefficient of Thermal Expansion of Composite Materials for Thin-Film Coatings

Xu Long [1,*], Tianxiong Su [1], Zubin Chen [1], Yutai Su [1,*] and Kim S. Siow [2]

[1] School of Mechanics, Civil Engineering and Architecture, Northwestern Polytechnical University, Xi'an 710072, China; stx@mail.nwpu.edu.cn (T.S.); zb.chen@mail.nwpu.edu.cn (Z.C.)
[2] Institute of Microengineering and Nanoelectronics, Universiti Kebangsaan Malaysia, Bangi 43600, Malaysia; kimsiow@ukm.edu.my
* Correspondence: xulong@nwpu.edu.cn (X.L.); suyutai@nwpu.edu.cn (Y.S.)

Citation: Long, X.; Su, T.; Chen, Z.; Su, Y.; Siow, K.S. Tunable Coefficient of Thermal Expansion of Composite Materials for Thin-Film Coatings. *Coatings* **2022**, *12*, 836. https://doi.org/10.3390/coatings12060836

Academic Editors: Adrian David and Sergey N. Grigoriev

Received: 3 May 2022
Accepted: 13 June 2022
Published: 15 June 2022

Abstract: In most engineering applications, the coefficients of thermal expansion (CTEs) of different materials in integrated structures are inconsistent, especially for the thin-film multilayered coatings. Therefore, mismatched thermal deformation is induced due to temperature variation, which leads to an extreme temperature gradient, stress concentration, and damage accumulation. Controlling the CTEs of materials can effectively eliminate the thermally induced stress within the layered structures and thus considerably improve the mechanical reliability and service life. In this paper, randomly distributed fibers are incorporated into the matrix material and thus utilized to tune the material CTE from the macroscopical viewpoint. To this end, finite element (FE) modeling is proposed for fiber-reinforced matrix composites. In order to overcome the challenges of creating numerical models at a mesoscale, the random distribution of fibers in three-dimensional space is realized by proposing a fiber growth algorithm with the control of the in-plane and out-of-plane angles of fibers. The homogenization method is adopted to facilitate the FE simulations by using the representative volume element (RVE) of composite materials. Periodic boundary conditions (PBC) are applied to realize the prediction of the equivalent CTE of macroscopic composite materials with randomly distributed fibers. In the established FE model, the random distribution of carbon fibers in the matrix makes it possible to tune the CTE of the composite material by considering the orientation of fibers in the matrix. The FE predictions show that the volume fraction of carbon fibers in the composite materials is found to be crucial to macroscopic CTE, but results in minor variations in Young's modulus and shear modulus. With the developed ABAQUS plug-in program, the proposed tuning method for CTE is promising to be standardized for industrial practice.

Keywords: fiber-reinforced composite; finite element simulation; coefficient of thermal expansion; representative volume element; periodic boundary conditions

1. Introduction

Fiber-reinforced matrix materials are composite materials formed by using fiber or carbon fiber fabric as reinforcement and resin, ceramic, metal, and cement as the matrix. Such composite materials have the advantages of convenient processing, high specific strength, and low relative density [1–3]. Such composite materials have been widely applied in various sectors of industries. As the core competency, the design and optimization of composite materials have been attracting more attention in scientific research and technological development. For fiber-reinforced matrix composites, it is important to understand the relationship between the macroscopic mechanical properties and the material microstructure. With the significant advance in computing capacity in recent years, numerical simulations can be performed for composite materials with increasingly complex microstructures, providing new ideas for the design of composite materials [4–6]. Based on the characteristics of fiber-reinforced matrix composites, it is difficult to accurately describe the mechanical properties of composites by traditional mechanical models at the macroscopic scale. Therefore, Panin [7] proposed the concept of micromechanics, which was also accepted by most scientific and technological workers. The advantage of this concept is that composite materials can be divided into multiple scale levels and analyzed individually. On the basis of Panin, Mishnaevsky [8] introduced the concept of mesoscopic continuum mechanics and concluded that this method also involves statistical principles and quantum mechanics.

With the rapid development of electronic packaging technology, chips are developing in the direction of miniaturization and high power. These all put forward higher requirements on the mechanical properties of packaging materials [9,10]. On this basis, some studies have been carried out on the influence of fiber materials on solder joints [11,12]. As the thermal effect continues to expand, the thermal failure problem of composite materials for thin-film coatings in the interconnect material in the package structure is also more serious. However, the current study on the coefficients of thermal expansion (CTEs) of composite materials for thin-film coatings still has great challenges.

Carbon-fiber-reinforced matrix composites have the advantages of high strength and isotropic elastic modulus, which are widely utilized in a wide range of applications such as aerospace and electronic packaging. However, fiber-reinforced materials are complex in their spatial configuration so far. Despite the extremely high randomness of the material at the mesoscopic level, it generally exhibits good property homogeneity at the macroscopic level. At appropriate geometrical scales, this localized inhomogeneity of the material can be manipulated and handled by mesoscale units, which are usually referred to as unit cells. In fact, there are a number of theoretical models and numerical methods for predicting the mechanical properties of carbon-fiber-reinforced composites. When performing calculations using mesomechanical models, Andriyana [13] considered not only the random distribution process of fibers but also the orientation of fibers. Wan and Takahashi [14] studied the effect of fiber aspect ratio on the tensile properties of materials by different modeling methods and obtained the fiber aspect ratio to achieve the best tensile properties based on the Mori–Tanaka model. However, it should be noted that the preparation process of carbon-fiber-reinforced composites is complicated, and there is an unaffordable number of variables involved in the preparation process. This leads to high research and development costs. Therefore, computer-aided design has been used to optimize the distribution and aspect ratio of fibers.

Mirkhalaf [15] carried out model design and nonlinear calculation using the reinforced matrix composite material module in the finite element (FE) software, and obtained the effects of different fiber volume fractions, fiber length–diameter ratios and orientation distributions on the macroscopic mechanical properties of composite material. Similarly, by using the Digimat-FE, Díaz [16] investigated the thermal conductivity of wood and Trzepieciński [17] completed the studies on the failure characteristics of composite materials. In addition, Chao et al. [18] found that chopped carbon fibers also have the characteristics of a negative thermal expansion coefficient in a certain direction. However, in the field of electronic packaging, there are few related studies on the contribution of carbon fibers to the mechanical properties of thin-film coating materials and the coefficient of thermal expansion of composites. Therefore, this paper investigates the tunable thermal expansion coefficient of composite materials for thin film coating.

In this study, a three-dimensional mesomechanical model of fiber-reinforced materials is developed by using the parameterized preprocess in ABAQUS. In this process, the application of the random distribution of carbon fibers in composites is realized through the secondary development script of the ABAQUS software, and the randomness of the direction of carbon fibers is included at the same time. In order to more deeply reveal the effect of carbon fiber microscale on the macroscopic mechanical properties of composites, the macroscopical coefficients of thermal expansion (CTEs) of carbon-fiber-reinforced matrix composites are calculated by using the representative volume element (RVE) method. Furthermore, the influence of the volume fraction of chopped carbon fibers on the macroscopic

CTE based on the RVE responses is explored, and it also provides a more reliable basis for the in-depth research and comprehensive application of thin-film coating materials. Accordingly, the study contents of this paper can be outlined as follows:

1. A simulation model of RVEs is established based on the theory of mesomechanics, and the process and difficulties of its establishment are described in detail.
2. The random distribution of the position and direction of the fiber material in the composite materials for thin-film coatings is realized by the calculation method combined with the relevant development script file of ABAQUS.
3. The close integration of periodic boundary condition theory and finite element simulation software is realized, and constraints are imposed on the corresponding nodes of the RVEs.
4. The macroscopic mechanical properties and CTE in RVEs are calculated based on the randomness of the fiber distribution.

Finally, the trend in the linear elastic properties and the CTE of the composites with fiber volume fraction are further analyzed, and it is found that there is a certain negative correlation, that is, with the increase in fiber volume fraction, the elastic modulus, shear modulus and CTE decrease gradually.

2. Modeling of Representative Volume Elements for Composite Materials

To perform efficient FE modeling for the fiber-reinforced matrix composites, the homogenization modeling approach is proposed on the basis of RVEs for analyzing the mesoscale mechanical properties of composites [19–22]. To represent the macroscopic material behavior and scale down the computational cost, periodic boundary conditions are applied on the RVE as well. Therefore, the equivalent CTE of the composites can be tuned through the adjustment of fiber properties [23–25]. In particular, the establishment of FE models involves the characteristics of the random distribution of fibers in the matrix and also considers the orientation of fibers inside the matrix, which is also a difficulty in establishing a mesomechanical model.

The mechanical properties of fiber-reinforced composites, including but not limited to the equivalent elastic properties, CTE, and volume fraction of fibers, can be investigated efficiently using RVE [26–28]. However, for composite materials with internally structured and randomly distributed fibers, it is important to reduce the computational cost by sensibly selecting numerical models, which should also be as large as possible by taking full account of the interaction between fibers and the matrix material. Thus, using a mesoscale RVE approach ensures both computational accuracy and efficiency. In order to adequately balance these contradictions, a mesomechanical approach is adopted in this paper to create three-dimensional numerical models of random fiber-reinforced matrix composites.

2.1. *Representative Volume Element Design Steps*

Traditional methods of designing fiber-reinforced matrix composites suffer from time-consuming design cycles and high manufacturing cost, which limit the efficiency of updating and iterating composites. This section realizes the rapid modeling of fiber-reinforced matrix composites through a Python program script file based on ABAQUS. On this basis, the influence of the material properties and geometric dimensions of the fibers and the matrix on the equivalent thermal expansion coefficient is studied.

In order to accurately simulate the distribution of three-dimensional fiber units in the matrix, the random distribution of fibers in the matrix is simulated according to the microstructure characteristics of fiber-reinforced matrix composites. Through the random generation algorithm, the initial coordinates of the fibers are determined, and the collective growth process of the fibers is determined by the placement algorithm. Based on this fiber placement algorithm, the accuracy of fiber placement can be significantly improved and the time required for model creation is reduced. When using the Python program script file to generate the model, the following issues should be paid attention to. The spatial position of the fiber inside the matrix usually includes fiber spatial coordinates, fiber radius, fiber length, fiber spatial orientation, and fiber-to-fiber distance judgment. Based on the above problems, using the Python program script based on ABAQUS, the creation process of RVEs of fiber-reinforced matrix composites in ABAQUS can be divided into the following five steps.

Step 1: Creation of a single fiber

As the fiber units are all solid units with the same radius and length, a fiber unit is first generated, and a specific number of fiber aggregates can be obtained by repeatedly repeating the operation. The resulting single fiber element is a solid element with a radius of 1.5 μm and a length of 10 μm.

Step 2: Generation process of fiber starting point coordinates

Generating a randomly distributed fiber usually includes the following parameters: the coordinates of the center of the fiber base, the radius of the fiber base, the total fiber length, and the fiber spatial orientation. Therefore, according to the uniform distribution of fibers in space, the function library of ABAQUS is used to generate a specific number of random numbers within a limited spatial range. Each group of random numbers represents the coordinates of the starting point of the fiber. The spatial distribution problem in the actual fiber formation process can be better simulated through this method, as shown in Figure 1.

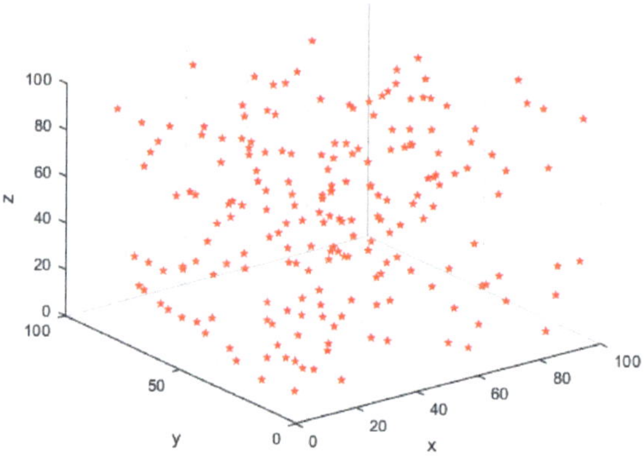

Figure 1. Three-dimensional coordinates of fibers randomly distributed in space.

Step 3: Generation process of fiber endpoint coordinates.

Assuming that the fiber is straight in the actual production process and the fiber does not bend during the growth process, the difference between the ending coordinate of the fiber and the starting coordinate of the fiber is the length and the spatial orientation value of the fiber. When a fiber rotates in three-dimensional space, it contains two parameters: the in-plane angle θ of the fiber, and the out-of-plane angle φ, as shown in Figure 2.

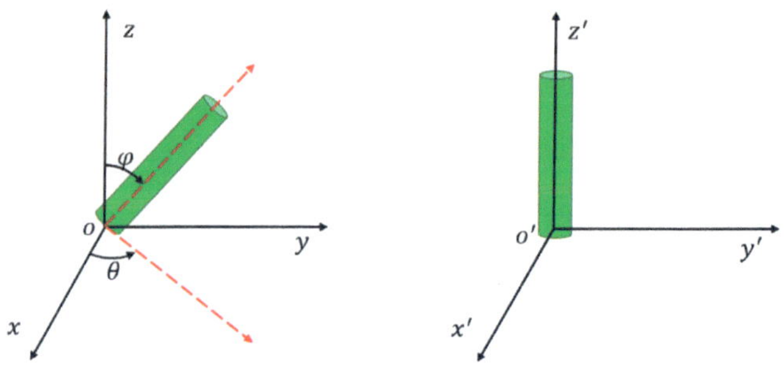

Figure 2. Angle parameter of fiber distribution in space.

The values of the spatial angles θ and φ of the fibers are randomly selected in the range of $[0, 2\pi]$ to ensure that the generated overall fibers have macroscopic isotropic properties in statistics. In order to further study the probability distribution of fiber orientation in space, a second-order tensor A_{ij} was introduced, which was decomposed into the product of eigenvalues and eigenvectors, and represented as a three-dimensional ellipse in space, as shown in Figure 3.

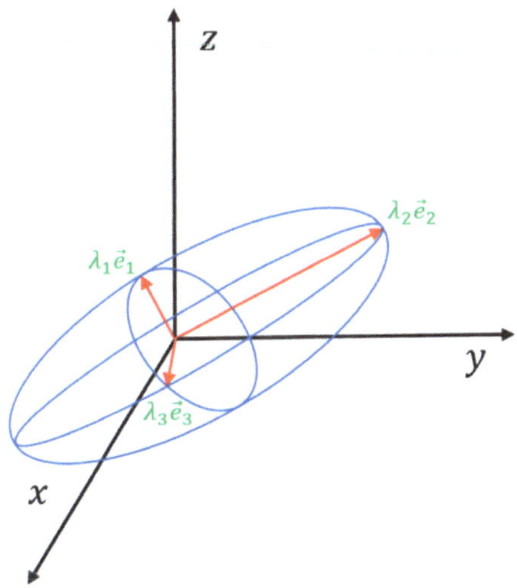

Figure 3. Definition of fiber orientation in three-dimensional space.

The direction vector formed from the center of the circle to any point on the ellipsoid can represent the distribution probability of the spatial fiber at this point. The direction distribution of each fiber is expressed by the angles θ and φ; then, the fiber distribution within a specific range can be expressed by

$$a_{ij} = \begin{bmatrix} sin^2\theta cos^2\theta & sin^2\theta sin\theta cos\varphi & sin\theta cos\theta cos\varphi \\ sin^2\theta sin\varphi cos\varphi & sin^2\theta sin^2\varphi & sin\varphi sin\theta cos\theta \\ cos\varphi sin\theta cos\theta & sin\varphi sin\theta cos\theta & sin^2\theta \end{bmatrix} \quad (1)$$

where a_{ij} is the orientation distribution, and through the above coordinate analysis and the analysis of fiber orientation in the fiber space, M and N can be used to represent the starting and end coordinates of the fiber; then, the starting point coordinates $M(M_x,\ M_y,\ M_z)$ and the end point $N(N_x,\ N_y,\ N_z)$ can be expressed as

$$\begin{aligned} M_x &= random.rand(a,b) \\ M_y &= random.rand(a,b) \\ M_z &= random.rand(a,b) \end{aligned} \quad (2)$$

$$\begin{aligned} N_x &= random.rand(a,b) + Lsin\varphi cos\theta \\ N_y &= random.rand(a,b) + Lsin\varphi sin\theta \\ N_z &= random.rand(a,b) + Lcos\varphi \end{aligned} \quad (3)$$

where θ is the in-plane angle, φ is the out-of-plane angle, L is the length of the fiber, a is the left boundary of the coordinate distribution, b is the right boundary of the coordinate, and *random* is the random distribution function inline in Python. According to Equations (2) and (3), the coordinates of the starting and end points of the fiber are determined, and the radius parameter R of the fiber is added to determine the relative position of the entire fiber inside the matrix.

Step 4: Interference determination of fibers based on spatial distance.

Since the starting point and end point coordinates of fibers are randomly generated, the phenomenon of fiber intersection inevitably occurs, and interference determination is required when the spatial position of the fiber is finally determined. When the phenomenon of fiber intersection occurs, the distribution law of fibers in the matrix cannot be well simulated, and also the intersected fibers lead to contact problems resulting in numerical divergence and also significantly increasing computational time. Therefore, in FE simulations, when the three-dimensional model of the fiber-reinforced matrix composite material is finally generated, it is also necessary to check the interference between the fibers.

The method of checking fiber interference is to determine the shortest distance between the centerlines of each fiber, which is usually greater than the diameter of the fiber's bottom surface. The center line of the fiber is a straight line in space, and the shortest distance between the two straight lines in space is the length of the common perpendicular. However, this paper needs to determine the shortest distance between the two lines in space, even if the distance of the common perpendicular is smaller than the fiber diameter. If the distance between the two ends of the fiber is ensured to be greater than the diameter, the fiber that meets the requirements may also be generated, as shown in Figure 4.

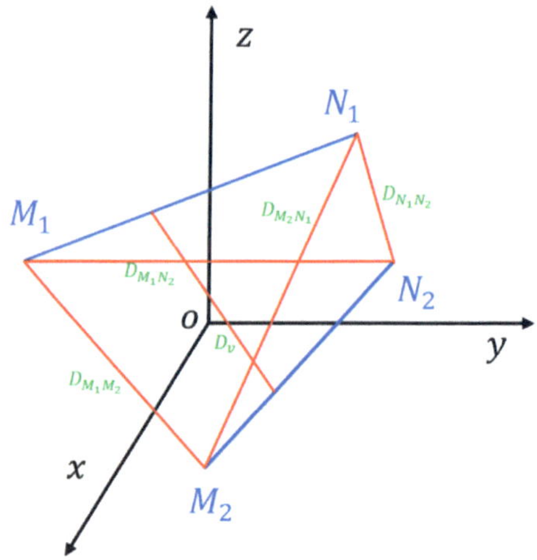

Figure 4. Determination of the distance between fibers.

When calculating the shortest distance between M_1N_1 and M_2N_2 of interspace line, it is necessary to obtain the common perpendicular line segment Dv of the different plane lines, and it is also necessary to calculate the lengths $D_{M_1M_2}$, $D_{M_1N_2}$, $D_{M_2N_1}$ and $D_{N_1N_2}$ of the lines between the four endpoints, and then compare with the diameter and length of the fiber, respectively, when D_v, $D_{M_1M_2}$, $D_{M_1N_2}$, $D_{M_2N_1}$ and $D_{N_1N_2}$ are smaller than the fiber diameter. It is determined that a fiber that meets the conditions has been generated, and the distance determination is calculated by

$$\begin{aligned} D_{M_1M_2} &= \sqrt{(M_{1x}-M_{2x})^2+(M_{1y}-M_{2y})^2+(M_{1z}-M_{2z})^2} \\ D_{M_1N_2} &= \sqrt{(M_{1x}-N_{2x})^2+(M_{1y}-N_{2y})^2+(M_{1z}-N_{2z})^2} \\ D_{M_2N_1} &= \sqrt{(M_{2x}-N_{1x})^2+(M_{2y}-N_{1y})^2+(M_{2z}-N_{1z})^2} \\ D_{N_1N_2} &= \sqrt{(N_{1x}-N_{2x})^2+(N_{1y}-N_{2y})^2+(N_{1z}-N_{2z})^2} \\ D_v &= \frac{\left|(\overrightarrow{M_1N_1}\times\overrightarrow{M_2N_2})\cdot\overrightarrow{M_1M_2}\right|}{\left|\overrightarrow{M_1N_1}\times\overrightarrow{M_2N_2}\right|} \end{aligned} \tag{4}$$

where $D_{M_1M_2}$ is the distance between points M_1 and M_2, $D_{M_1N_2}$ is the distance between points M_1 and N_2, $D_{M_2N_1}$ is the distance between points M_2 and N_1, $D_{N_1N_2}$ is the distance between points N_1 and N_2, and D_v is the distance between vertical line segments of interspace line M_1N_1 and M_2N_2.

Step 5: Generating fiber solid model based on ABAQUS python function library.

Through the script interface of ABAQUS, all the fiber space parameters that meet the requirements are processed into data and further reconstructed in ABAQUS, as shown in Figure 5. The reconstructed functions used can call the function library included in ABAQUS.

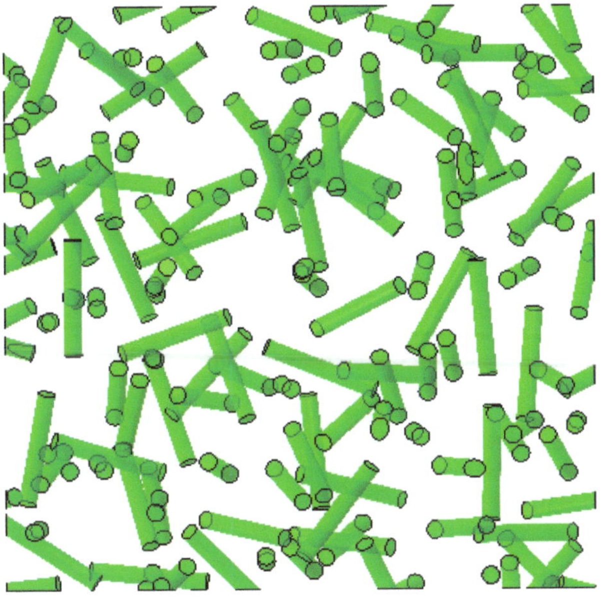

Figure 5. Randomly distributed fibers.

2.2. Periodic Boundary Conditions for the RVE

Fiber-reinforced matrix composites have local randomness. When using RVEs to carry out finite element numerical simulation of reinforced matrix composites, it is necessary to ensure the continuity between elements. There are three realization methods in numerical simulation to achieve this consistency on the boundary, given surface force, given displacement, and imposed periodic boundary conditions. However, for the model adopted in this paper, giving a surface force and displacement is unrealistic, so RVEs are applied in the periodic boundary conditions (PBC) [29]. In addition, Chen [30] applied each of the above boundary conditions using commercial FE software when investigating the equivalent mechanical properties of 2D porous materials and compared the final finite element analysis results with experiments. After comparison, it is found that when the PBC is applied to the representative element, the obtained FE analysis results are the closest to the actual properties of the material, and when the boundary conditions of a given surface force and a given displacement are used, the obtained finite element analysis results are obtained. There is often a large deviation in the numerical value between the analysis results and the real mechanical properties of the material. In addition, Chen et al. [30] also mentioned that if the real material model to be simulated has irregular shapes on the boundary surface, the real situation of the material can be more accurately predicted by using PBC by eliminating interfacial effects on irregular hole boundaries, as shown in Figure 6. Therefore, PBC is applied to the RVE boundary according to the research object and the above research results.

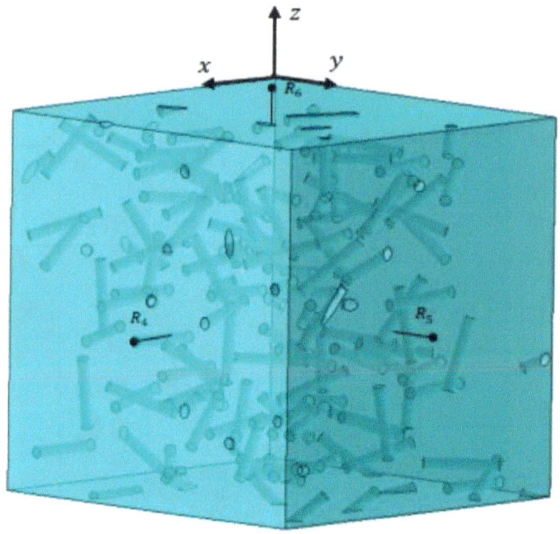

Figure 6. Schematic diagram of PBC for an RVE.

Based on the application method of periodic boundary conditions, deformation control can be performed on the boundary of RVEs so that the opposite surfaces of the elements have deformation compatibility. During the finite element simulation, each outer surface of the RVE has consistent deformation and force so that the entire fiber-reinforced matrix composite model conforms to the basic assumptions of continuum mechanics. The finite element analysis has laid the foundation. Suquet [31] conducted a theoretical derivation on the theory of periodic boundary conditions, and the specific derivation process can use the displacement field defined as Equation (5). In addition, Xia [32] completed further verification and application on this basis.

$$U_i = \bar{\varepsilon}_{ik} x_k + u_i^* \tag{5}$$

where $\bar{\varepsilon}_{ik}$ is the average strain of the representative volume unit, x_k is any position inside the representative volume unit, and u_i^* is the periodic displacement correction of the boundary position. The deformation coordination condition of Equation (5), u_i^*, is an unknown quantity in the deformation process, and is only related to the global load on the RVE, so this periodic displacement field cannot be applied to the actual operation process. In the fiber-reinforced matrix composite structure mentioned in this paper, the boundary surfaces of this structure are parallel so that the periodic displacement field can be written as

$$\begin{aligned} U_i^{j+} &= \bar{\varepsilon}_{ik} x_k^{j+} + u_i^* \\ U_i^{j-} &= \bar{\varepsilon}_{ik} x_k^{j-} + u_i^* \end{aligned} \tag{6}$$

where j^+ and j^- represent the positive and negative directions of the representative volume unit, respectively. Since in the periodic boundary condition, the value of u_i^* is the same on the two opposite surfaces, subtracting the displacement expressions on the left and right boundaries yields

$$U_i^{j+} - U_i^{j-} = \bar{\varepsilon}_{ik}(x_k^{j+} - x_k^{j-}) = \bar{\varepsilon}_{ik} \Delta x_k^j \tag{7}$$

where Δx_k^j is the relative displacement value of the relative boundary. For a specific $\bar{\varepsilon}_{ik}$, the displacement change on the right side of the above Equation is a constant, so the above equation can be rewritten as

$$u_i^{j+}(x,y,z) - u_i^{j-}(x,y,z) = c_i^j (i,j = 1,2,3) \tag{8}$$

By observing Equation (8), it can be found that the improved formula does not contain the correction amount of the periodic displacement. ABAQUS is used to impose periodic boundary conditions when the RVE is used for the simulation calculation. According to the above theoretical derivation process, more MPC point constraint equations are added to achieve stress–strain continuum conditions.

3. Equivalent CTE of Representative Volume Elements

This paper introduces three types of boundary conditions in finite element analysis. Among them, the displacement-based analysis conditions are natural boundary conditions, and the uniqueness of the solution is solved in the analysis process, but such boundary conditions cannot accurately describe the RVE in the stress–strain field induced during the deformation process. When applying periodic boundary conditions in ABAQUS, the grid nodes on the opposite surface can be used to control the application of MPC multipoint constraint equations. Most literature only mentions the mathematical expression of periodic boundary conditions and does not mention the realization process in ABAQUS finite element analysis, which is a considerable challenge for numerical analysis. This section mainly expounds on the theoretical basis and implementation methods of applying periodic boundary conditions in ABAQUS and then performs corresponding operations on the corresponding nodes, edge nodes, and vertices on the parallel plane of the RVE, as shown in Figure 7.

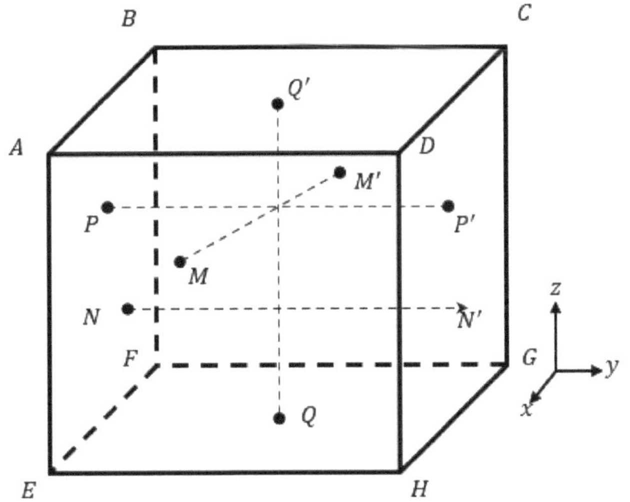

Figure 7. Boundary conditions for multipoint constraint equations.

In this paper, in the finite element software ABAQUS, the RVE is generated by the Python scripts. In addition, to ensure the consistency of the deformation of the symmetrical points and surfaces of the RVE, it is necessary to use the method of Python script files. The equivalent linear elastic mechanical properties of fiber-reinforced matrix composites will be studied later. The modeling process of the RVE of the fiber-reinforced matrix composite material in the finite element software was introduced in detail above. This paper adopts the finite element model method to calculate the equivalent linear elastic mechanical properties of the RVE. The method for calculating the equivalent linear elastic mechanical properties of RVEs in this paper is based on the ABAQUS plug-in program.

In this paper, the thermal expansion coefficient of the fiber-reinforced base needs to be regulated, so for the convenience of research, the elastic material properties of the selected two-phase materials are given in Tables 1 and 2.

Table 1. Material properties of reinforced fibers.

Elastic (GPa)	E_1 **174**	E_2 **174**	E_3 **9.6**	G_{12} **70.4**	G_{13} **3.7**	G_{23} **3.7**	ν_{12} **0.234**	ν_{13} **0.273**	ν_{23} **0.273**
CTE ($\times 10^{-6}$/K)		α_{11} −0.07044			α_{22} −0.07044			α_{33} 10.4956	

Table 2. Mechanical properties of the matrix material.

Material Name	**Elastic Modulus (GPa)**	**Poisson's Ratio**	**α ($\times 10^{-6}$/K)**
SAC305	20	0.4	24

Fiber is a transversely isotropic material. When calculating with ABAQUS, it is necessary to assign a local coordinate system to the material, and the thermal expansion coefficient of the fiber is also related to the direction.

According to the data in Table 2, fibers are regarded as transversely isotropic materials, each of which has an elastic axis of symmetry. The elastic constitutive relationship at any two positions with this axis as the axis of symmetry is all the same. Therefore, when a plane is perpendicular to the elastic symmetry axis, all directions in the plane are exactly symmetrical to the elastic symmetry axis, and the elastic constitutive relationship is the same. In transversely isotropic materials, the in-plane constitutive relation has in-plane isotropy. The typical representative in nature is the flaky distribution of rocks. In this paper, the elastic constitutive relation of the fibers in the isotropic plane is the same.

The elastic symmetry axis of the fiber is the radial direction of the fiber, so the direction of the Z axis is taken as the elastic symmetry axis of the fiber, and the coordinate axes X and Y establish an isotropic symmetry plane, as shown in Figure 8. In the isotropic symmetry plane, when the X and Y coordinates are exchanged, the elastic constitutive relation of the material does not change. In order to satisfy this condition, the elastic constitutive Jacobian matrix coefficient must satisfy

$$\begin{aligned} c_{11} &= c_{22} \\ c_{13} &= c_{23} \\ c_{55} &= c_{66} \end{aligned} \tag{9}$$

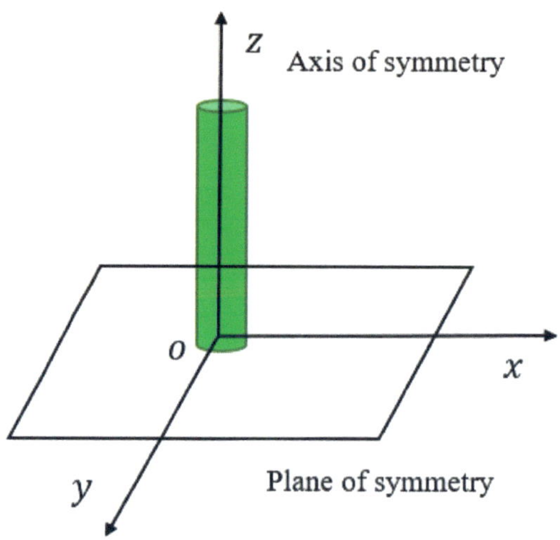

Figure 8. Elastic constitutive properties of fibers.

Additionally, according to the invariant principle of the shear stress–strain relationship given by:

$$c_{44} = \frac{1}{2}(c_{11} - c_{12}) \tag{10}$$

Through the above analysis, the elastic material properties of fibers can be described only by five independent elastic constants, and the elastic matrix of fibers can be written as

$$\begin{bmatrix} c_{11} & c_{12} & c_{13} & 0 & 0 & 0 \\ c_{12} & c_{11} & c_{13} & 0 & 0 & 0 \\ c_{13} & c_{13} & c_{13} & 0 & 0 & 0 \\ 0 & 0 & 0 & \frac{1}{2}(c_{11} - c_{12}) & 0 & 0 \\ 0 & 0 & 0 & 0 & c_{55} & 0 \\ 0 & 0 & 0 & 0 & 0 & c_{55} \end{bmatrix} \tag{11}$$

When using elastic parameters in ABAQUS, it also needs to be converted into a stiffness matrix. Invert Equation (11) to obtain Equation (12):

$$\begin{Bmatrix} \varepsilon_x \\ \varepsilon_y \\ \varepsilon_z \\ \gamma_{xy} \\ \gamma_{yz} \\ \gamma_{zx} \end{Bmatrix} = \begin{bmatrix} \frac{1}{E} & -\frac{\nu}{E} & -\frac{\nu'}{E} & 0 & 0 & 0 \\ -\frac{\nu}{E} & \frac{1}{E} & -\frac{\nu'}{E} & 0 & 0 & 0 \\ -\frac{\nu'}{E'} & -\frac{\nu'}{E'} & \frac{1}{E'} & 0 & 0 & 0 \\ 0 & 0 & 0 & \frac{1}{G} & 0 & 0 \\ 0 & 0 & 0 & 0 & \frac{1}{G'} & 0 \\ 0 & 0 & 0 & 0 & 0 & \frac{1}{G'} \end{bmatrix} \begin{Bmatrix} \sigma_x \\ \sigma_y \\ \sigma_z \\ \tau_{xy} \\ \tau_{yz} \\ \tau_{zx} \end{Bmatrix} \tag{12}$$

For transversely isotropic materials, the elastic axis of symmetry for each fiber needs to be defined in ABAQUS, as shown in Figure 9. However, the direction of each fiber is random. When the number of fibers reaches dozens or hundreds, how to accurately apply an elastic axis of symmetry to each fiber is a technical problem that needs to be solved urgently.

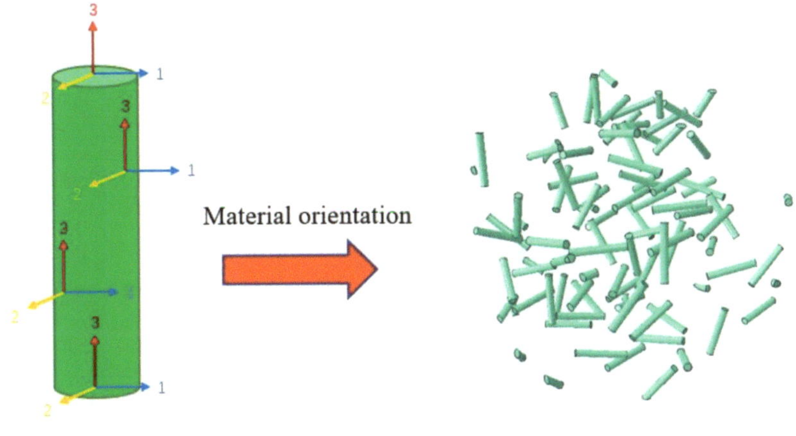

Figure 9. Material orientation of fibers.

In this paper, the Python script based on ABAQUS preprocessing is used to realize the assignment of the local coordinate system to all random fibers. In ABAQUS, a local coordinate system needs to be applied to each fiber first. For fibers, the application of the local coordinate system is often related to the bottom surface and bottom surface circumference of the fiber. The bottom surface is selected to define the isotropic surface of the fiber, and the bottom surface circumference is selected to define the elastic axis of symmetry of the fiber. Since the bottom surface and bottom surface circumference of each fiber are different, it is necessary to use the preprocessing script to define the set of bottom surface and bottom surface circumference each time random fibers are generated to lay the foundation for the subsequent automatic addition of material directions, as shown in Figure 10. Figure 10 highlights the local coordinate direction of each fiber in the model, and uses 1, 2, and 3 displayed in different colors to represent the three axes of x, y, and z of the spatial coordinate system.

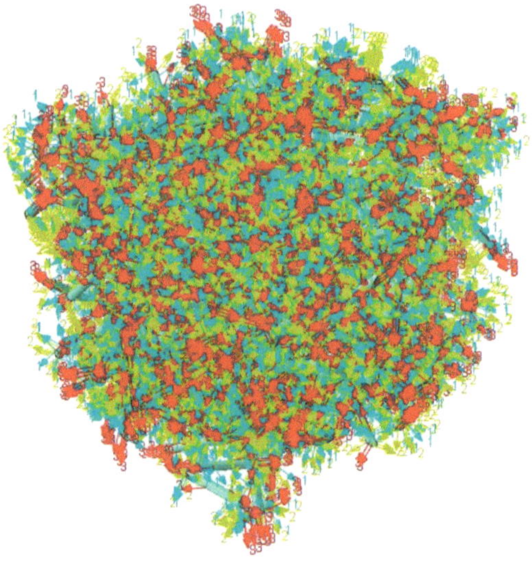

Figure 10. Schematic diagram of fiber local coordinate system definition.

Similar to the method by Chao [18], the fibers used in this paper have a negative thermal expansion coefficient on the isotropic plane and a positive thermal expansion coefficient in the radial direction of the fibers, which means that when the temperature increases, the shrinkage ratio of the fibers in the circumferential direction will be greater than the elongation ratio in the radial direction so that the overall control of the equivalent thermal expansion coefficient can be achieved. For the electronic packaging thin-film coating materials studied in this paper, the calculation is performed assuming that the base material is an isotropic material. Based on the above analysis process, this paper has studied the equivalent mechanical properties and thermal expansion coefficients of fiber-reinforced matrix composites with different volume fractions.

4. Result Discussion

In this paper, the generation method of fibers is to control the random generation number of the coordinates of the fiber starting point. During each iteration of the calculation, the different volume fraction of fibers can be easily generated by modifying the number of fibers in the model. The tetrahedral mesh is used in the generated finite element model, and the element type is C3D10, as shown in Table 3.

According to Table 3, as the number of iterations increases, the number of fibers increases steadily, increasing the volume fraction of fibers. Due to the strong randomness of the spatial distribution of fibers, the generated finite element model needs to be divided into many meshes. When the fiber volume fraction is close to 2%, it reaches more than 400,000 meshes, still in the mesh density. When more extensive, this presents a higher challenge to finite element calculation. Through finite element calculation, the data shown in Table 4 are obtained.

Table 3. Fiber integration in the generated finite element models.

Fiber Numbers	Volume Fraction	Mesh Number	Computing Time (s)
50	0.2356	149,560	430
100	0.4799	205,991	885
150	0.6646	255,566	1142
200	0.8695	302,956	1537
250	1.1039	320,125	1914
300	1.2512	348,464	2085
350	1.3916	348,464	2423
400	1.6425	367,355	2751
450	1.8191	398,205	3005
500	1.9704	412,653	3363

Table 4. Equivalent linear elastic mechanical properties of fiber-reinforced matrix composites.

Volume	0.23%	0.47%	0.66%	0.86%	1.10%	1.25%	1.39%	1.64%	1.81%	1.97%
E_1(GPa)	19.989	19.971	19.968	19.953	19.942	19.935	19.926	19.909	19.902	19.886
E_2(GPa)	19.988	19.974	19.965	19.955	19.942	19.933	19.926	19.913	19.899	19.895
E_3(GPa)	19.987	19.972	19.967	19.954	19.939	19.936	19.925	19.910	19.901	19.890
G_{12}(GPa)	7.819	7.804	7.784	7.771	7.765	7.751	7.737	7.722	7.712	7.705
G_{13}(GPa)	7.814	7.799	7.787	7.779	7.761	7.757	7.748	7.733	7.720	7.712
G_{23}(GPa)	7.812	7.803	7.787	7.776	7.769	7.756	7.753	7.736	7.726	7.721
ν_{12}	0.277	0.277	0.276	0.276	0.275	0.275	0.274	0.274	0.274	0.273
ν_{13}	0.278	0.277	0.277	0.276	0.276	0.275	0.274	0.274	0.274	0.273
ν_{23}	0.277	0.276	0.276	0.276	0.276	0.275	0.275	0.274	0.273	0.273

According to Table 4, the fibers act as a transversely isotropic material, but the phenomenon that makes the material properties of composites different in different directions is due to the fibers being distributed in the matrix according to the characteristics of random starting positions and random fiber orientations. In addition, with the continuous increase in the fiber volume fraction, the equivalent linear elastic mechanical properties of the composite show a decreasing trend. The values of E_1, E_2, and E_3 are very close, and it can be found that it can be approximated as a linear decrease. For the equivalent shear modulus of fiber-reinforced matrix materials, the numerical differences in the three planes are not large, and it can be approximately considered that with the increase in fiber volume fraction, the equivalent shear modulus approximately exhibits a linearly decreasing trend. The equivalent Poisson's ratios in the three principal planes do not change much and can be considered consistent. Therefore, the above-generated fiber-reinforced matrix composites can be regarded as isotropic materials, and the calculated stress distributions are shown in Figures 11 and 12.

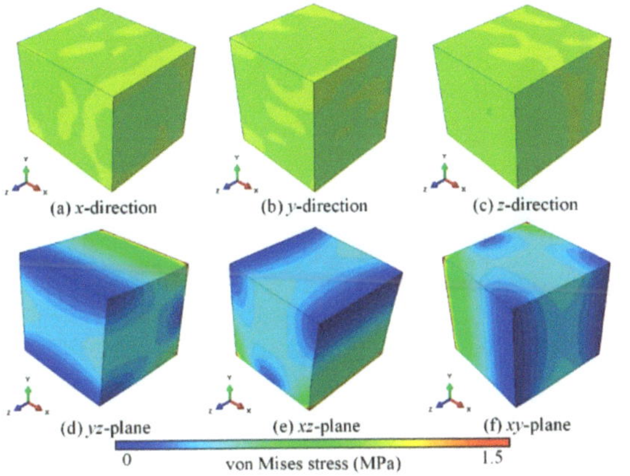

Figure 11. Equivalent linear elastic properties of RVEs.

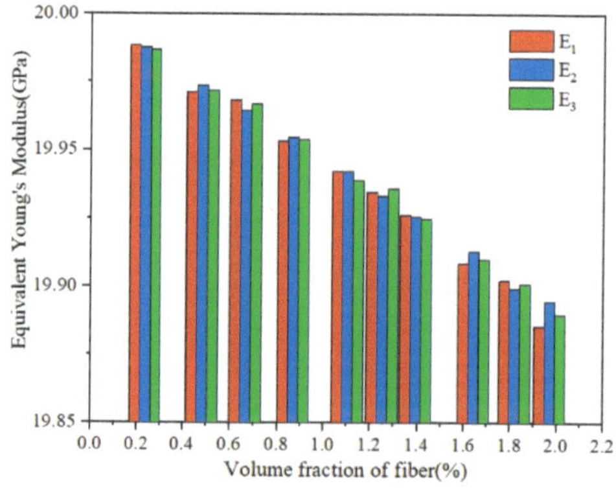

Figure 12. Variation trend in equivalent Young's modulus.

As shown in Figure 12, the equivalent moduli E_1, E_2, E_3 in the three planes are very close. When the volume fraction is 0, it is an isotropic material with an elastic modulus of 20 GPa. At the same time, it can be found that the values of E_1, E_2, and E_3 decrease linearly with the fiber volume fraction. After fitting the Origin data, Equation (13) can be obtained.

$$\begin{aligned} E_1 &= 20 - 0.055 \times Volum\_F \\ E_2 &= 20 - 0.053 \times Volum\_F \\ E_3 &= 20 - 0.054 \times Volum\_F \end{aligned} \tag{13}$$

where E_1, E_2, E_3 are three different equivalent linear elastic moduli, and $Volum\_F$ is the fiber volume fraction.

As shown in Figure 13, the equivalent shear modulus of the fibers in the three planes has approximately the same value. With the continuous increase in the fiber volume fraction, the equivalent shear modulus shows a linearly decreasing trend. The data fitting can obtain Equation (14):

$$\begin{aligned} G_{12} &= 7.833 - 0.066 \times Volum\_F \\ G_{13} &= 7.828 - 0.058 \times Volum\_F \\ G_{23} &= 7.825 - 0.054 \times Volum\_F \end{aligned} \tag{14}$$

where G_{12}, G_{13}, G_{23} are three different equivalent shear moduli, and $Volum\_F$ is the fiber volume fraction. In addition, the macroscopic thermal expansion coefficient of fiber-reinforced matrix composites needs to be analyzed, and the results are shown in Figure 14.

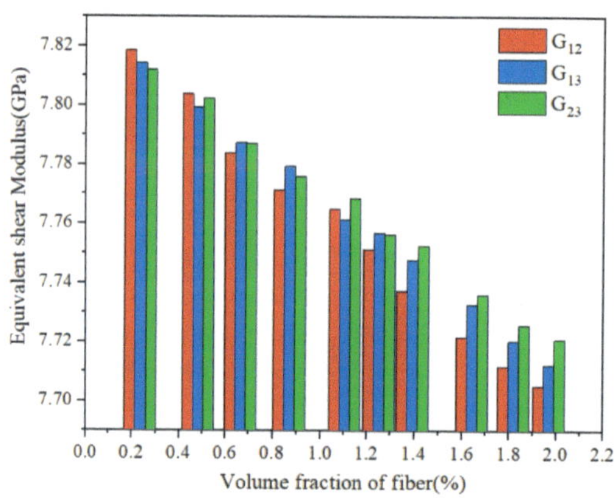

Figure 13. Variation trend in equivalent shear modulus.

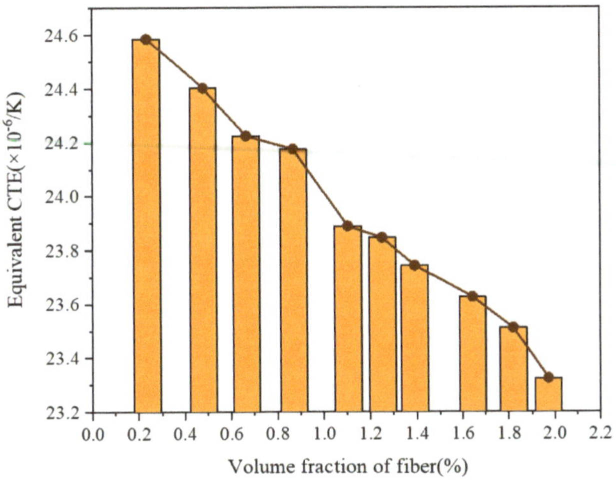

Figure 14. Trend in equivalent CTE with the variation in volume fraction of fibers.

It can be seen from that as the fiber content continues to increase, the equivalent thermal expansion coefficient of the representative volume unit continues to decrease. It can be seen from the numerical value that the overall thermal expansion coefficient decreases by a considerable amount, which is limited by the calculation cost and calculation accuracy. The highest content of fiber volume fraction discussed in this paper is close to 2%, and the empirical Equation (15) is obtained by linear fitting of Origin data:

$$\alpha_{eq} = 2.47267 \times 10^{-5} - 6.94244 \times 10^{-7} \times Volum\_F \tag{15}$$

where α_{eq} is the equivalent thermal expansion coefficient, $Volum\_F$ is the fiber volume fraction.

According to the change trends in the above calculation results, further analysis is completed and it is found that the reason for the equivalent Young's modulus and shear modulus showing linear decreasing trends as the volume fraction of the fibers increases is due to the linear elastic properties of the carbon fiber and base matrix material. Moreover, the different CTEs of the carbon fiber and matrix material cause the CTE of the thin-film coating material to change with the carbon fiber volume fraction. Additionally, the CTE of carbon fiber is much smaller than that of the matrix material, which also causes the CTE of the thin-film coating material to decrease with the increase in the carbon fiber volume fraction.

5. Conclusions

This paper mainly introduced the study on regulating the coefficient of thermal expansion of composite materials for thin-film coatings used in the field of electronic packaging. Based on the advantages of convenience, efficiency, and versatility of the commercial finite element software ABAQUS, numerical simulations of fiber-reinforced matrix composites were achieved by utilizing the representative volume element. Additionally, according to the theory of mesomechanics, periodic mesh division and the application of periodic boundary conditions were realized on the finite element model. Moreover, the equivalent linear elastic mechanical properties of composites were further studied, including the equivalent Young's modulus, equivalent shear modulus, and the variation law of the effective thermal expansion coefficient with the fiber volume fraction.

When the volume fraction of fibers accounts for 2% of the matrix, the Young's modulus of the fiber-reinforced matrix composite is adjusted to 19.886 GPa, the shear modulus is 7.7 GPa, and the equivalent thermal expansion coefficient is 23.324 ppm/K. The control effect of linear elastic mechanical properties is noticeable. This paper mainly has the following two conclusions:

1. The mesomechanical modeling of fiber-reinforced matrix composites is complex, involving the random distribution and random orientation of fibers. In this paper, the parametric modeling method based on Python program script files was used to improve modeling efficiency significantly.
2. Through the finite element analysis of the fiber-reinforced matrix composite material model, it was found that the equivalent thermal expansion coefficient of the material can be effectively reduced with the continuous increase in the fiber volume fraction.

Author Contributions: Conceptualization, X.L. and Y.S.; investigation, T.S. and Z.C.; writing—original draft preparation, X.L. and Z.C.; writing—review and editing, T.S., K.S.S. and Y.S.; funding acquisition, X.L.; supervision, X.L. All authors have read and agreed to the published version of the manuscript.

Funding: This work was supported by the National Natural Science Foundation of China (No. 52175148), the Natural Science Foundation of Shaanxi Province (No. 2021KW-25), the Open Cooperation Innovation Fund of Xi'an Modern Chemistry Research Institute (No. SYJJ20210409), and the Fundamental Research Funds for the Central Universities (No. 3102018ZY015).

Institutional Review Board Statement: Not applicable.

Informed Consent Statement: Not applicable.

Data Availability Statement: Not applicable.

Conflicts of Interest: The authors declare no conflict of interest.

References

1. Zeng, Q.; Yu, A.; Lu, G.M. Interfacial interactions and structure of polyurethane intercalated nanocomposite. *Nanotechnology* **2005**, *16*, 2757. [CrossRef]
2. Hashimoto, M.; Okabe, T.; Sasayama, T.; Matsutani, H.; Nishikawa, M. Prediction of tensile strength of discontinuous carbon fiber/polypropylene composite with fiber orientation distribution. *Compos. Part A Appl. Sci. Manuf.* **2012**, *43*, 1791–1799. [CrossRef]
3. Kwon, D.J.; Jang, Y.J.; Choi, H.H.; Kim, K.; Kim, G.H.; Kong, J.; Nam, S.Y. Impacts of thermoplastics content on mechanical properties of continuous fiber-reinforced thermoplastic composites. *Compos. Part B Eng.* **2021**, *216*, 108859. [CrossRef]
4. Lee, J.Y.; Baljon, A.; Loring, R.F.; Panagiotopoulos, A.Z. Simulation of polymer melt intercalation in layered nanocomposites. *J. Chem. Phys.* **1998**, *109*, 10321–10330. [CrossRef]
5. Smith, G.D.; Bedrov, D.; Li, L.W.; Byutner, O. A molecular dynamics simulation study of the viscoelastic properties of polymer nanocomposites. *J. Chem. Phys.* **2002**, *117*, 9478–9489. [CrossRef]
6. Thompson, R.B.; Ginzburg, V.V.; Matsen, M.W.; Balazs, A.C. Block copolymer-directed assembly of nanoparticles: Forming mesoscopically ordered hybrid materials. *Macromolecules* **2002**, *35*, 1060–1071. [CrossRef]
7. Panin, V.E. *Physical Mesomechanics of Heterogeneous Media and Computer-Aided Design of Materials*; Cambridge International Science Publishing: Cambridge, UK, 1998.
8. Mishnaevsky, L.L., Jr.; Schmauder, S. Contimuum mesomechanical finite element modeling in materials development: A state-of-the-art review. *Adv. Mech.* **2002**, *54*, 49.
9. Long, X.; Guo, Y.; Chang, X.; Su, Y.; Shi, H.; Huang, T.; Tu, B.; Wu, Y. Effect of Temperature on the Fatigue Damage of SAC305 Solder. In Proceedings of the 2021 22nd International Conference on Electronic Packaging Technology (ICEPT), Xiamen, China, 14–17 September 2021; pp. 1–4.
10. Long, X.; Su, T.; Chang, C.; Huang, J.; Chang, X.; Su, Y.; Shi, H.; Huang, T.; Wu, Y. Mechanical Behavior of Lead-Free Solder at High Temperatures and High Strain Rates. In Proceedings of the 2021 22nd International Conference on Electronic Packaging Technology (ICEPT), Xiamen, China, 14–17 September 2021; pp. 1–4.
11. Andriyana, A.; Billon, N.; Silva, L. Mechanical response of a short fiber-reinforced thermoplastic: Experimental investigation and continuum mechanical modeling. *Eur. J. Mech. A Solids* **2010**, *29*, 1065–1077. [CrossRef]
12. Wan, Y.; Takahashi, J. Tensile properties and aspect ratio simulation of transversely isotropic discontinuous carbon fiber reinforced thermoplastics. *Compos. Sci. Technol.* **2016**, *137*, 167–176. [CrossRef]
13. Mirkhalaf, S.M.; Eggels, E.H.; van Beurden, T.J.H.; Larsson, F.; Fagerström, M. A finite element based orientation averaging method for predicting elastic properties of short fiber reinforced composites. *Compos. Part B Eng.* **2020**, *202*, 108388. [CrossRef]
14. Díaz, A.R.; Flores, E.I.S.; Yanez, S.J.; Vasco, D.A.; Pina, J.C.; Guzmán, C.F. Multiscale modeling of the thermal conductivity of wood and its application to cross-laminated timber. *Int. J. Therm. Sci.* **2019**, *144*, 79–92. [CrossRef]

15. Trzepieciński, T.; Ryzińska, G.; Biglar, M.; Gromada, M. Modelling of multilayer actuator layers by homogenisation technique using Digimat software. *Ceram. Int.* **2016**, *43*, 3259–3266. [CrossRef]
16. Chao, X.J.; Qi, L.H.; Tian, W.L.; Lu, Y.F.; Li, H.J. Potential of porous pyrolytic carbon for producing zero thermal expansion coefficient composites: A multi-scale numerical evaluation. *Compos. Struct.* **2020**, *235*, 111819. [CrossRef]
17. Miller, W.; Hook, P.B.; Smith, C.W.; Wang, X.; Evans, K.E. The manufacture and characterisation of a novel, low modulus, negative Poisson's ratio composite. *Compos. Sci. Technol.* **2009**, *69*, 651–655. [CrossRef]
18. Sun, Z.; Shan, Z.D.; Shao, T.M.; Li, J.H.; Wu, X.C. A multiscale modeling for predicting the thermal expansion behaviors of 3D C/SiC composites considering porosity and fiber volume fraction. *Ceram. Int.* **2021**, *47*, 7925–7936. [CrossRef]
19. Wei, K.; Peng, Y.; Wang, K.Y.; Duan, S.Y.; Yang, X.J.; Wen, W.B. Three dimensional lightweight lattice structures with large positive, zero and negative thermal expansion. *Compos. Struct.* **2018**, *188*, 287–296. [CrossRef]
20. Liu, H.; Zeng, D.; Li, Y.; Jiang, L.Y. Development of RVE-embedded solid elements model for predicting effective elastic constants of discontinuous fiber reinforced composites. *Mech. Mater.* **2016**, *93*, 109–123. [CrossRef]
21. Kobayashi, S.; Shimpo, T.; Goto, K. Microscopic damage behavior in carbon fiber reinforced plastic laminates for a high accuracy antenna in a satellite under cyclic thermal loading. *Adv. Compos. Mater.* **2019**, *28*, 259–269. [CrossRef]
22. Hao, X.L.; Zhou, H.Y.; Mu, B.S.; Chen, L.; Guo, Q.; Yi, X.; Sun, L.C.; Wang, Q.W.; Ou, R.X. Effects of fiber geometry and orientation distribution on the anisotropy of mechanical properties, creep behavior, and thermal expansion of natural fiber/HDPE composites. *Compos. Part B Eng.* **2020**, *185*, 107778. [CrossRef]
23. Gu, J.P.; Leng, J.S.; Sun, H.Y.; Zeng, H.; Cai, Z.B. Thermomechanical constitutive modeling of fiber reinforced shape memory polymer composites based on thermodynamics with internal state variables. *Mech. Mater.* **2019**, *130*, 9–19. [CrossRef]
24. Sun, Q.P.; Zhou, G.W.; Meng, Z.X.; Jain, M.; Su, X.M. An integrated computational materials engineering framework to analyze the failure behaviors of carbon fiber reinforced polymer composites for lightweight vehicle applications. *Compos. Sci. Technol.* **2021**, *202*, 108560. [CrossRef]
25. Chen, L.L.; Gu, B.Q.; Zhou, J.F.; Tao, J.H. Study of the effectiveness of the RVEs for random short fiber reinforced elastomer composites. *Fibers Polym.* **2019**, *20*, 1467–1479. [CrossRef]
26. Naili, C.; Doghri, I.; Kanit, T.; Sukiman, M.S.; Aissa-Berraies, A.; Imad, A. Short fiber reinforced composites: Unbiased full-field evaluation of various homogenization methods in elasticity. *Compos. Sci. Technol.* **2020**, *187*, 107942. [CrossRef]
27. Garoz, D.; Gilabert, F.A.; Sevenois, R.D.B.; Spronk, S.W.F.; Van Paepegem, W. Consistent application of periodic boundary conditions in implicit and explicit finite element simulations of damage in composites. *Compos. Part B Eng.* **2019**, *168*, 254–266. [CrossRef]
28. Chen, L.; Tao, X.M.; Choy, C.L. Mechanical analysis of 3-D braided composites by the finite multiphase element method. *Compos. Sci. Technol.* **1999**, *59*, 2383–2391. [CrossRef]
29. Suquet, P. Elements of homogenization for inelastic solid mechanics. In *Homogenization Techniques for Composite Media;* Springer: Berlin/Heidelberg, Germany, 1987.
30. Xia, Z.H.; Zhang, Y.F.; Ellyin, F. A unified periodical boundary conditions for representative volume elements of composites and applications. *Int. J. Solids Struct.* **2003**, *40*, 1907–1921. [CrossRef]
31. Xia, Z.H.; Zhou, C.W.; Yong, Q.L.; Wang, X.W. On selection of repeated unit cell model and application of unified periodic boundary conditions in micro-mechanical analysis of composites. *Int. J. Solids Struct.* **2006**, *43*, 266–278. [CrossRef]
32. Xia, W.X.; Oterkus, E.; Oterkus, S. Ordinary state-based peridynamic homogenization of periodic micro-structured materials. *Theor. Appl. Fract. Mech.* **2021**, *113*, 102960. [CrossRef]

coatings

Article

Viscoelastic Simulation of Stress and Warpage for Memory Chip 3D-Stacked Package

Xiyou Wang, Sicheng Cao, Guangsheng Lu and Daoguo Yang *

College of Mechanical and Electrical Engineering, Guilin University of Electronic Science Technology, Guilin 541004, China
* Correspondence: daoguo_yang@163.com

Abstract: Three-dimensional-stacked packaging technology is widely used in memory chip packaging, which can greatly increase the utilization ratio of the packaging area. However, problems with the reliability of 3D-stacked packaging are also becoming more and more serious. In this paper, first, a dynamic mechanical analyzer is used to obtain the EMC viscoelasticity parameters. Then, the influence trend of different factors, such as EMC, die bond material and chip, on the performance of the memory chip 3D-stacked packaging under a fixed temperature cyclic loading condition is explored by the FE method.

Keywords: 3D-stacked packaging; finite element analysis (FEA); epoxy molding compound (EMC); viscoelasticity; memory chip

Citation: Wang, X.; Cao, S.; Lu, G.; Yang, D. Viscoelastic Simulation of Stress and Warpage for Memory Chip 3D-Stacked Package. *Coatings* **2022**, *12*, 1976. https://doi.org/10.3390/coatings12121976

Academic Editor: Bohayra Mortazavi

Received: 7 November 2022
Accepted: 13 December 2022
Published: 16 December 2022

1. Introduction

With the continuous development of electronic information technology, higher requirements are put forward for the ability of information storage, which promotes packaging technology to develop towards miniaturization and high integration. The appearance of three-dimensional (3D) packaging technology greatly increases the utilization ratio of the packaging area, which is widely used in memory chip packaging. However, because of the material characteristics of plastic packaging, problems such as warpage, delamination and wire sweep in the packaging process that reduce the reliability of chips are becoming increasingly serious. To improve package reliability, the rational design of 3D chip packaging structure optimization is key.

The material properties of epoxy molding compounds (EMCs) change greatly during packaging processes [1–3]. In particular, EMCs can become viscoelastic under high-temperature processes. Therefore, the analysis of EMC materials that are subject to high-temperature processes such as a post-mold cure must take into account their potentially viscoelastic behavior [4]. The material viscoelastic constitutive model is closely related to time, temperature and cure degree. Therefore, selecting a reasonable material constitutive model is very important to correctly predict package warpage and stress.

The Finite Element Method (FEM) is widely employed for the prediction of the mechanical properties of packaging materials. For example, the influences of the structural and material parameters on the package warpage of PoP (Package on Package) and TSOP (Thin Small Outline Package) packaging were studied, in which the linear elastic constitutive model was adopted [5,6].

Kim et al. used FEM to compare and analyze the influence of the linear elastic constitutive model and the viscoelastic constitutive model on the warpage of BGA packaging. Compared with the experimental results, it was found that the viscoelastic constitutive model can be better used to predict the warpage of packaging than the linear elastic constitutive model [7].

A significant number of papers have employed FEM to study the mechanical properties of IC packages, such as TSOP [8], SiP [9] and BGA [10], in which a viscoelastic-based material model is adopted to account for the time- and temperature-dependent behavior of the EMC.

To use the viscoelastic constitutive model, material parameter measurements are highly important. The dynamic mechanical analyzer (DMA) test with the time–temperature superposition principle is the mainstream technology to obtain the material viscoelastic constitutive model of the EMC. The tensile storage modulus and loss modulus of EMCs under different frequency and temperatures conditions can be obtained by DMA experiments [11–15]; the main curves of storage modulus and loss modulus can be obtained using the time–temperature superposition principle, and the time–temperature conversion factor can be obtained by using WLF mode [16].

FEM, in conjunction with the viscoelastic constitutive model, has been found to be an effective approach to exploring the mechanical properties of IC packaging. However, few studies have focused on 3D-stacked chip packages, especially those in the working state. This paper studies the 3D-stacked packaging structure of memory chips. Considering the viscoelasticity of the EMC, this paper intends to explore the influence trend of different factors, such as EMC, die bond material and chip, on the 3D-stacked package under the temperature cyclic loading condition.

2. Finite Element Analysis

2.1. Geometric Model

The 3D-stacked package memory chip is involved the following components: stacked chips, a substrate, a die bond material and an EMC [17]. The package structure and sizes are provided in Figure 1 and Table 1. Note that the bonding wire is ignored in the FE model.

(a) (b) (c)

Figure 1. Three-dimensional-stacked package memory chip structure. (**a**) Package outline; (**b**) laser decap package sample and its left-top corner detail picture; (**c**) package cross-section schematic (chips 1 to 4 are memory chips, and chip 5 is a main-control chip).

Table 1. Packaging size and material parameters.

Item	Material	Size (mm)	E (MPa)	v	A (ppm/°C)
EMC	EMC	12 × 10 × 0.9	Test	0.3	9.38, T < Tg 38.68, T > Tg
Chip	Si	10 × 8.0 × 0.1 10 × 7.8 × 0.1 10 × 7.6 × 0.1 10 × 7.4 × 0.1 4 × 1.0 × 0.1	131,000	0.3	2.8
DA	DAF	10 × 8.0 × 0.02 10 × 7.8 × 0.02 10 × 7.6 × 0.02 10 × 7.4 × 0.02 4 × 1.0 × 0.02	640@25 °C 140@150 °C	0.4	48.00

Table 1. *Cont.*

Item	Material	Size (mm)	E (MPa)	v	A (ppm/°C)
substrate	BT	12 × 10 × 0.2	X, Y: 26,000 Z: 11,000	XZ, YZ: 0.39 XY: 0.11	X, Y: 15 Z: 50

Note: X/Y/Z is material parameters in X/Y/Z direction.

2.2. *Packaging Material Parameters*

The material parameters of the memory chip stacked package listed in Table 1 were largely obtained from the material supplier's documents and the relevant literature [18,19]. The modulus of the EMC was obtained through experiments in the following section. The EMC, chip and adhesive materials are all isotropic, while the substrate is orthotropic.

2.3. *EMC Viscoelasticity*

2.3.1. Experimental Instruments and Samples

The dynamic mechanical analyzer (DMA Q800, TA, Newcastle, DE, USA) was used to obtain the EMC's viscoelasticity parameters. EMC samples were obtained from Hitachi Chemical Co., Ltd., Suzhou, China, and belong to the CEL-9240HF series. In the preparation process, EMC pellets were first pressed and formed at 175 °C through a transfer molding machine and then fully cured at 175 °C for 8 h. A sample size of 42 mm × 8.5 mm × 1.5 mm was used, as shown in Figure 2.

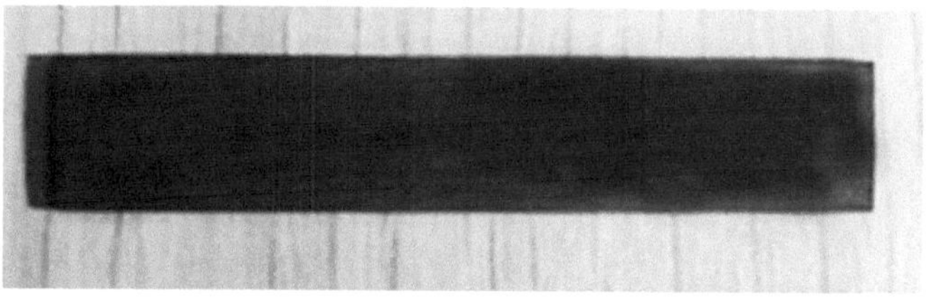

Figure 2. EMC samples for DMA test.

2.3.2. Experimental Data and Data Processing

A single cantilever clamp in DMA Q800 was used for experiments, as shown in Figure 3. The DMA detection mode was adjusted to the multi-frequency strain scanning mode with an amplitude of 10 μm. The starting temperature of the test was set to 25 °C, which was then increased to 245 °C at a rate of 0.4 °C/min. The set frequency of the test is shown in Table 2.

Figure 3. Single cantilever clamp in DMA.

Table 2. Excitation frequency table for tensile mode DMA experiments.

No.	1	2	3	4	5	6	7	8	9	10	11	12	13
Frequency (Hz)	0.1	0.18	0.32	0.56	1	1.8	3.2	5.6	10	17.8	31.6	56	100

2.3.3. Experimental Data

In this experiment, the sample data directly obtained by the DMA Q800 include the storage modulus and loss modulus. The test data curves are shown in Figures 4 and 5, respectively, which all show a strong temperature dependence.

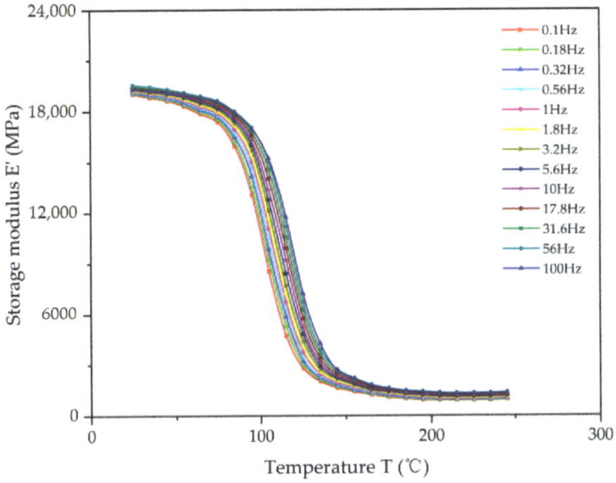

Figure 4. Storage modulus vs. temperature at different frequencies.

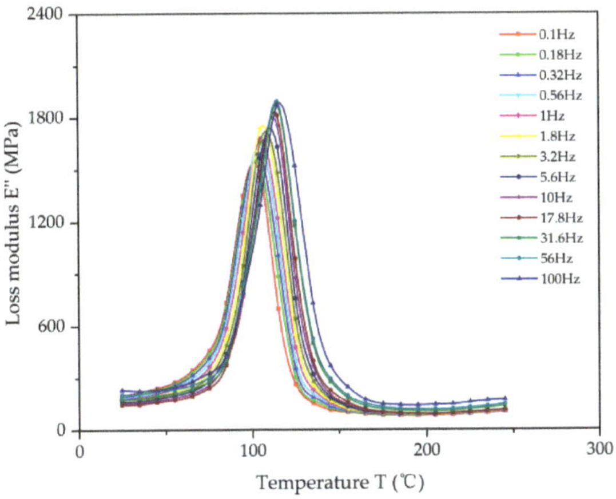

Figure 5. Loss modulus vs. temperature at different frequencies.

The storage modulus E′ is used to describe the ability of viscoelastic materials to store energy in the elastic deformation part, and its essence is Young's modulus. It can be seen from Figure 4 that, when the sample is in the low-temperature stage, it is in a glassy state, which shows a certain hardness and brittleness. The storage modulus decreases with increasing temperature, up until the point when the sample transitions to the viscoelastic state, which causes the modulus to decrease significantly. When the sample enters into the rubber state, the modulus decreases at a slower rate and tends to be stable.

The loss modulus E″ is used to describe the ability of viscoelastic materials to lose energy as it is converted into heat when irreversible viscous deformation occurs in the material, thus reflecting the viscous nature of the material. It can be seen from Figure 5 that, with the continuous increase in test frequency, the loss modulus tends to increase, and the peak value moves slowly towards the high-temperature range. In general, the smaller the loss modulus is, the smaller the damping loss factor of the material is, and its mechanical performance is much closer to that of an ideal elastomer.

In order to construct a master curve of storage modulus with a wider frequency domain, first, the curves of storage modulus and temperature obtained by experiments in Figure 4 are transformed into the correlation between storage modulus E′ and frequency w, as shown in Figure 6. It can be seen from the figure that the glass transition temperature Tg is about 105 °C. When the sample is in the low- and high-temperature range, the storage modulus remains the same with the increase in frequency. Changes in the modulus are mainly distributed in the temperature range of 85−135 °C.

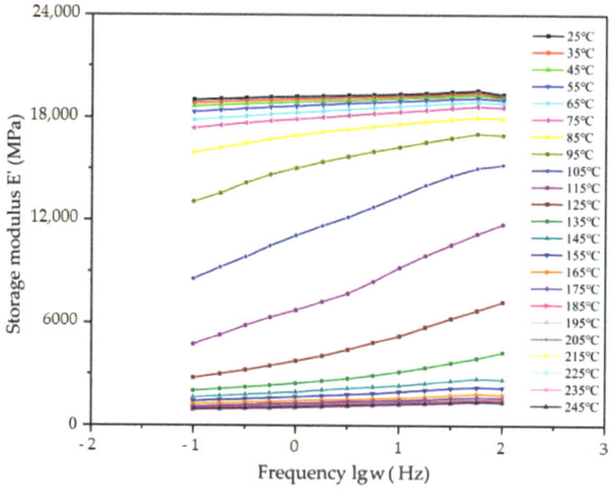

Figure 6. Storage modulus vs. frequency under different temperatures.

Based on the time–temperature superposition principle, the glass transition temperature Tg is selected as the reference temperature. The storage modulus curve at 105 °C shown in Figure 6 remains unchanged, while the storage modulus curves at other temperatures are moved and superimposed on both sides of the reference curve along the horizontal coordinate axis direction. The main storage modulus curve shown in Figure 7 can thus be obtained. Taking the reference curve at 105 °C as the zero-displacement point, the displacement of each curve in the horizontal direction is the displacement factor $\log\alpha_T$, as shown in Figure 8. A nonlinear fitting is performed on the scatter plot of Figure 8 using equations WLF [16], Equation (1), that is, the empirical constant of the material is obtained,

$$loga_T = \frac{-C_1 T - T_r}{C_2 + T - T_r} \quad (1)$$

where C_1 and C_2 are empirical constants, T is the test temperature and T_r is the reference temperature.

The main curve of the storage modulus obtained by the time–temperature superposition principle cannot be directly applied to FEA, and it needs to be converted into the corresponding shear modulus and bulk modulus according to Equations (2) and (3) [20].

$$G'(t) = \frac{E'(t)}{2(1+v)} \tag{2}$$

$$K'(t) = \frac{E'(t)}{3(1-2v)} \tag{3}$$

where $G'(t)$ is the shear modulus, $K'(t)$ is the bulk modulus, $E'(t)$ is the storage modulus, t is the time and v is Poisson's ratio.

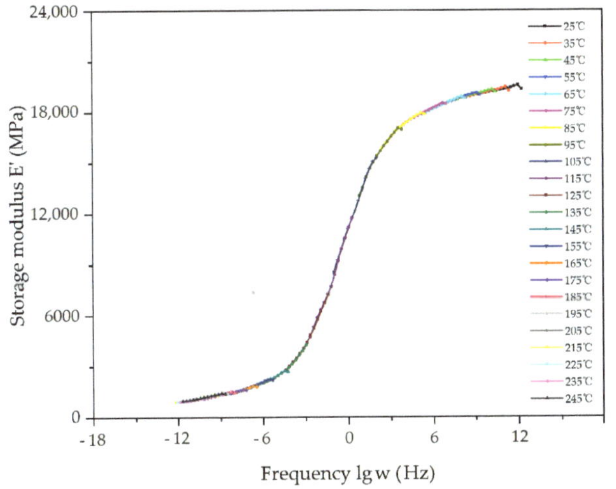

Figure 7. The master curve of storage modulus for EMC sample.

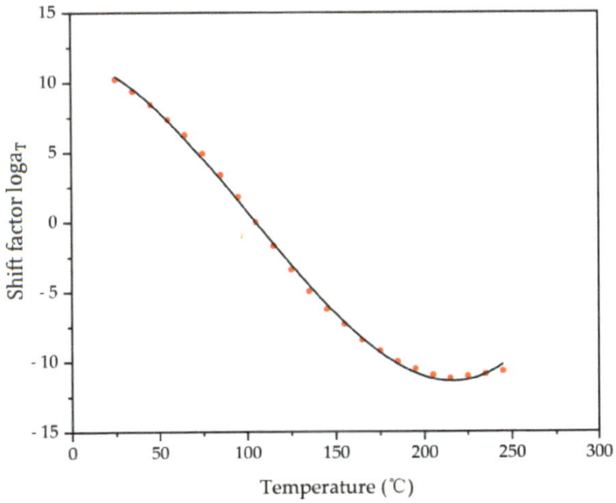

Figure 8. Shift factor scatter plot.

The shear modulus and bulk modulus can be expressed by the generalized Maxwell model [21], shown as Equations (4) and (5), respectively,

$$K'(\omega) = K_\infty + \sum_{i=1}^{n} K_i \frac{\omega^2 \tau_i^2}{1 + \omega^2 \tau_i^2} \tag{4}$$

$$G'(\omega) = G_\infty + \sum_{i=1}^{n} G_i \frac{\omega^2 \tau_i^2}{1 + \omega^2 \tau_i^2} \tag{5}$$

where K_∞ is the bulk modulus in the rubber state, G_∞ is the shear modulus in the rubber state and τ_i is the relaxation time.

According to the obtained shear modulus curve and bulk modulus curve, the shear relaxation coefficient G_i and bulk relaxation coefficient K_i applied in the finite element simulation can be obtained by nonlinear fitting. The relevant material parameters are summarized in Table 3.

Table 3. Viscoelastic material parameters of epoxy molding compound.

No.	Relaxation Time τ_i (s)	Shear Relaxation Coefficient G_i	Volume Relaxation Coefficient K_i
τ1	1.00×10^{-8}	239.32	518.54
τ2	1.00×10^{-6}	272.59	590.62
τ3	1.00×10^{-4}	354.33	767.71
τ4	1.00×10^{-3}	349.54	757.34
τ5	1.00×10^{-2}	513.60	1112.8
τ6	1.00×10^{-1}	966.29	2093.6
τ7	1.00×10^{0}	853.56	1849.4
τ8	1.00×10^{1}	1106.5	2397.4
τ9	1.00×10^{2}	726.77	1574.6
τ10	1.00×10^{3}	560.19	1213.8
τ11	1.00×10^{4}	475.34	1029.7
τ12	1.00×10^{6}	305.22	662.19
τ13	1.00×10^{8}	87.513	183.44
τ14	1.00×10^{9}	150.87	342.30

2.4. FE Mold

2.4.1. FE Mesh

Using FEA software MSC.MARC (version 16.0) to study the mechanical property of packaging, the geometric modeling is carried out according to the memory chip 3D-stacked package structure, and the corresponding material properties are given in Table 1. The mesh generation is divided by plane stretching into hexahedral mesh elements, and the mesh elements of the contact part are refined. The properties are set as three-dimensional solid elements, as shown in Figure 9. The total number of grid cells reached 572885.

Figure 9. Package material properties and FE grid.

2.4.2. Boundary Conditions and Load

Since the structure of the package is not completely symmetrical, in the analysis process, the nodes near the center area at the bottom of the substrate are selected for complete displacement constraint (U = v = w = 0), the adjacent nodes are fixed in the Y direction (v = 0) and Z direction (w = 0), respectively, and the remaining nodes remain free.

For the thermal stress simulations, an initial ambient temperature of 25 °C was set as the zero-stress state, and a loading temperature cycle of −55° C−125 °C with a cycle time of 3360 s for 5 cycles was used. The temperature cycle loading curve is shown in Figure 10, where the temperature rises and drops at a rate of 10 °C/min, and the constant temperature is maintained for 10 min after reaching the high- and low-temperature stages.

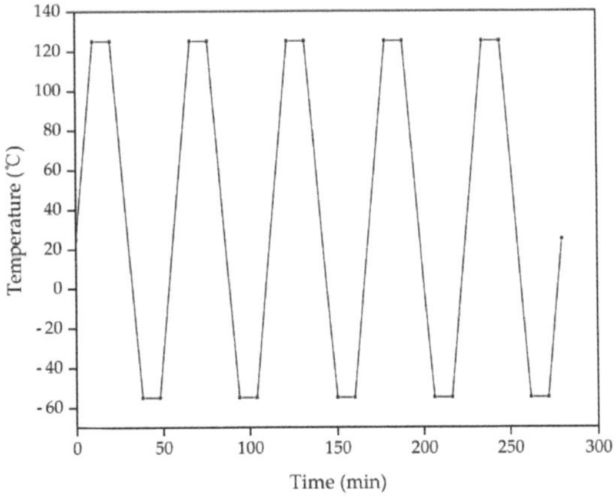

Figure 10. Temperature cyclic loading curve.

3. Results and Discussion

3.1. Analysis of Stress Simulation Results

Young's modulus is usually used to describe the anti-deformation ability of solid materials. Since the glass transition temperature of the EMC samples obtained in the previous experiments is only 105 °C, the EMC transitions from the glass state to the rubber state when the temperature rises above 105 °C during the temperature cycle simulation process. As such, the modulus decreases greatly, resulting in large warpage of the package at the high-temperature stage, reaching 21.84 μm. This occurs at the top corner of the EMC material of the package, as shown in Figure 11a.

Although the warpage at high temperature is large, the stress relaxation of the viscoelastic EMC material also occurs such that the stress phenomenon inside the package is effectively alleviated. A maximum stress value of only 61.58 MPa is reached. When the constant temperature stage ends and the temperature drops below 105 °C, the modulus begins to rise sharply and the deformation ability of the EMC becomes worse, resulting in less deformation in the low-temperature stage. The internal stress of the package also begins to increase with the increase in the EMC modulus. The maximum stress obtained in the low-temperature stage is 75.63 MPa, which occurs at the corner of the interface between the lowest chip and the adhesive material. The maximum stress change in the cycle is shown in Figure 12.

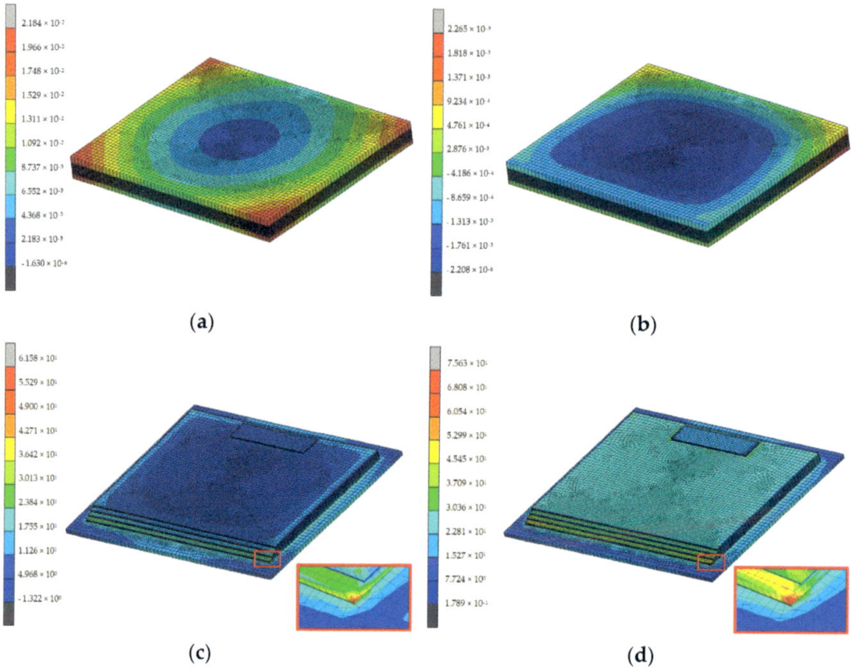

Figure 11. Package deformation and stress cloud picture. (**a**) Package deformation cloud picture at high temperature; (**b**) package deformation cloud picture at low temperature; (**c**) package stress cloud picture at high temperature; (**d**) package stress cloud picture at low temperature.

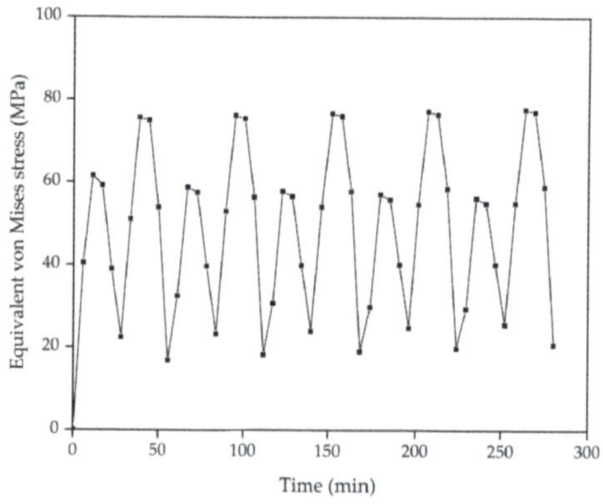

Figure 12. Change in maximum stress in load cycle.

The maximum stress of chip-1 to chip-5 is 75.63, 59.60, 49.44, 46.94 and 26.60 MPa in increasing chip order, which shows a decreasing trend from bottom to top, as shown in Figure 13. It can be seen from the stress cloud picture of each chip that the maximum stress always occurs at the interface between the chip stacking step and the adhesive material and is far less than the breaking stress limit of 6–7 GPa of silicon material. This proves that the packaging structure meets the process standard from the perspective of stress analysis. Therefore, the following sections of this paper mainly focus on the optimization design of the package warpage.

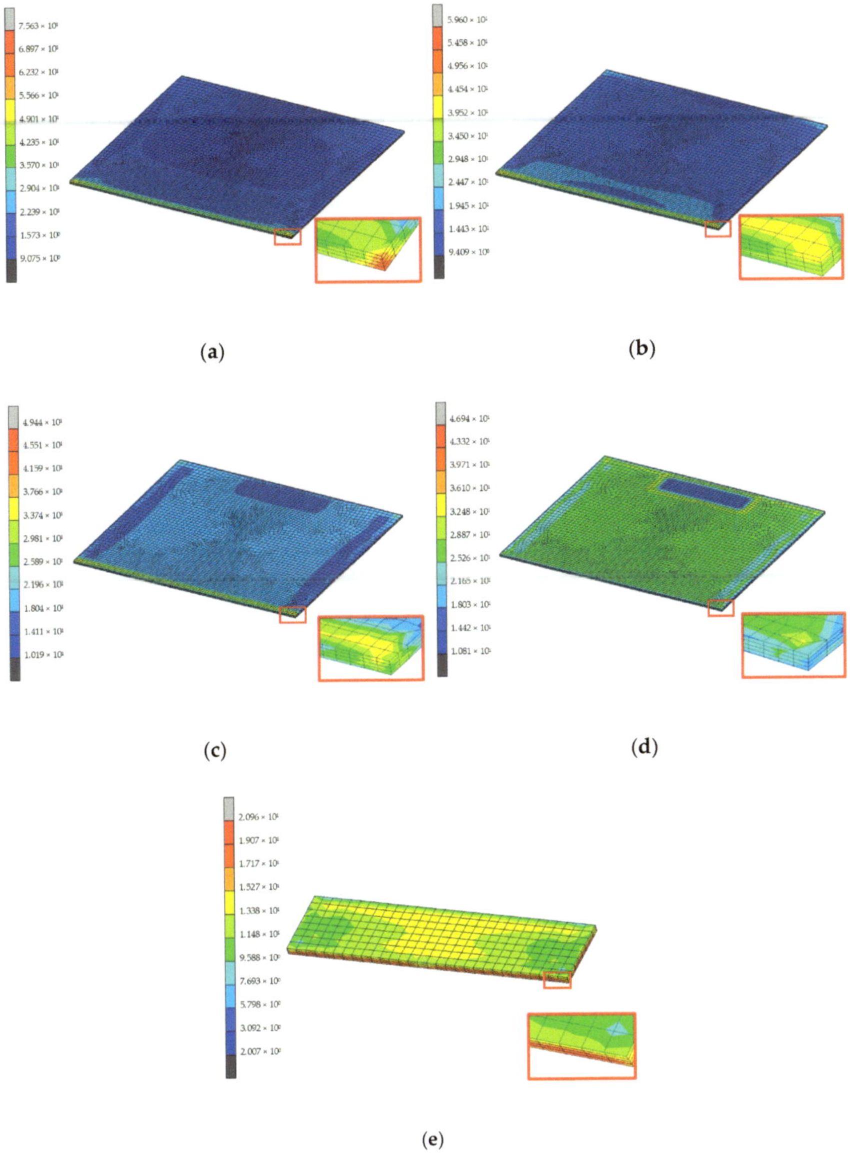

Figure 13. Stress cloud picture of each layer of stacked chips. (**a**) Chip-1 stress cloud picture; (**b**) Chip-2 stress cloud picture; (**c**) Chip-3 stress cloud picture; (**d**) Chip-4 stress cloud picture; (**e**) Chip-5 stress cloud picture.

3.2. Analysis of Warpage Simulation Results

The mismatch of the thermal expansion coefficient between the materials of the package is the root cause of warpage [22]. In addition to this, changes in the internal structure and material properties can influence warpage. It is generally required that the warpage should not exceed 50 μm [18] in the height direction. In this study, the package warpage value under a temperature cycle is selected as an evaluating indicator in order to explore the influence of packaging structure on package warpage. In addition, different applications of memory chips often lead to changes in the substrate structure. It should be noted that the study mainly analyzes the general rules of the influence of packaging materials on the warpage of the package and does not consider the influence of the substrate.

3.2.1. Influence of EMC Thickness on Warpage

From the perspective of the packaging structure, the increase in the thickness of the EMC helps to improve warpage in the production process and extend the distance between the internal chip and the external environment, and thus reduce the possibility of moisture diffusion corrosion. However, the increase in the thickness of the EMC will also lead to poor heat dissipation performance of the chip. The design of 3D packaging must therefore aim to find the balance between internal junction temperature and package warpage. The influence curve of the EMC thickness on the package warpage is shown in Figure 14.

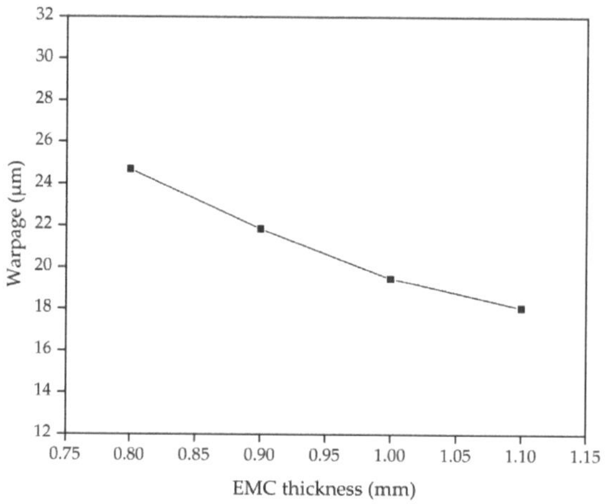

Figure 14. Influence of EMC thickness on warpage.

The results show that with the increase in the thickness of EMC material, the warpage value of the package decreases and reaches a minimum value of 18.08 μm when the thickness is 1.1 mm. The changing range is more than 5 μm, and therefore the thickness of the EMC material can be regarded as one of the main influencing factors. At a thickness of 0.8 mm EMC material, the top chip is only 0.2 mm away from the EMC surface, and the bonding wire loop height here is 150 μm. Considering that subsequent laser marking requires a 50 μm safe distance, a 0.2 mm EMC thickness is regarded as the limit. In combination with the above analysis, in order to effectively reduce the reliability problems caused by packaging warpage, the thickness of the EMC can be appropriately increased within the allowable range of the structure.

3.2.2. Influence of Die Bond Material on Warpage

In the packaging process of three-dimensional memory chips, DAF (die attach film) is usually used for the die bond [23], with a typical thickness of 5–30 μm [24]. In this section, the influence of DAF thickness and the coefficient of thermal expansion (CTE) on warpage is investigated. The influence curve of DAF on the package warpage is shown in Figure 15.

There are five layers of DAF inside the package, where every 5 μm increment in the DAF thickness causes a 25 μm change in the package thickness. The simulation results show that the warpage value of the package increases gradually with the increase in DAF thickness, where the maximum value of 23.39 μm is obtained when a DAF thickness is increased to 30 μm. In addition, the increase in the CTE of DAF will lead to a smaller increase in warpage, with a maximum warpage value of 22.17 μm obtained at 80 ppm/°C of DAF CTE. The increasing amplitude is 0.7 μm with different DAF CTE. Therefore, the thickness of DAF should be controlled as much as possible, and a low CTE should be selected.

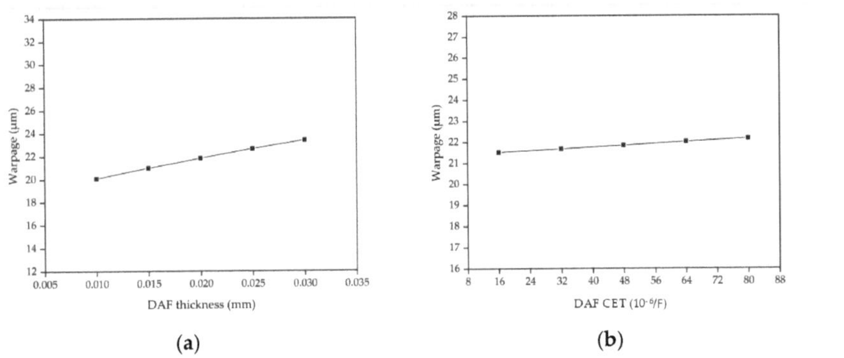

Figure 15. Influence of DAF on warpage. (a) Influence of DAF thickness on warpage; (b) influence of DAF CET on warpage.

3.2.3. Influence of Die Thickness on Warpage

The wafer original thickness is usually 750 μm, which will be reduced to 320 μm, 230 μm and even 50 μm when packaging. The circuit itself only occupies fewer than 5 μm, but with the decrease in wafer thickness, its strength also decreases. For the 3D-stacked packaging of memory chips, the number of internal integrated chips determines the capacity of information storage. However, the control of chip thickness is not only related to the packaging process but also needs to consider the influence of deformation and thermal stress during the packaging process.

There are five layers of chips inside the package, where every 10 μm increment in chip thickness leads to a 25 μm change in the package thickness. The simulation results show that with the increase in the chip thickness, the package warpage shows a weak downward trend and obtains a minimum value of 20.89 μm when chip thickness decreases to 130 μm, as shown in Figure 16. The material of the chip is monocrystalline silicon, which has a large Young's modulus and a thermal expansion coefficient much lower than that of all other packaging materials. The increase in the chip thickness leads to an increase in volume occupied by the silicon material in the package, which affects the deformation ability of the package. Therefore, the chip thickness can be appropriately increased to resist the influence of package warpage.

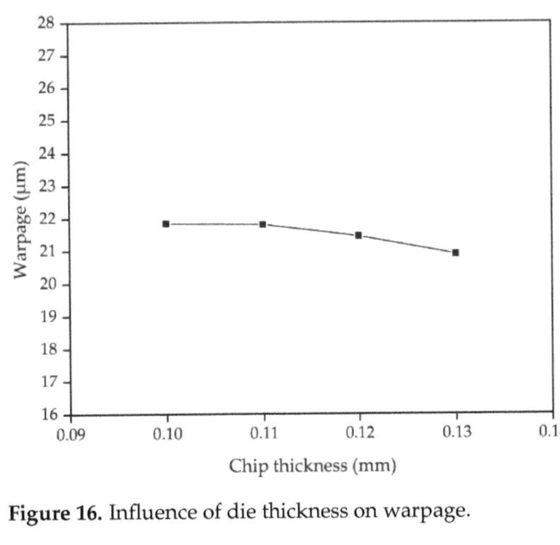

Figure 16. Influence of die thickness on warpage.

4. Conclusions

In this study, using the FE method and considering the viscoelasticity of the EMC, this paper uses DMA to obtain the EMC viscoelasticity parameters and then uses the FE method to explore the influence trend of different factors, such as EMC, die bond material and chip, on the performance of the packaging under the temperature cyclic loading condition. The results show the following: (1) in the temperature cycle simulation process, when the temperature rises above 105 °C, the EMC transitions from the glass state to the rubber state, and the modulus decreases greatly, resulting in large package warpage up to 21.84 μm in the high-temperature stage; (2) although the warpage at high temperature is large, the stress relaxation of the viscoelastic EMC material also occurs at this time. This effectively alleviates the stress inside the package to a value far less than the breaking stress limit of silicon material; (3) different package structures could lead to different package warpage, and the influence of EMC thickness on warpage is much greater than the die and die bond material. To effectively reduce the reliability problems caused by packaging warpage, the thickness of the EMC can be appropriately increased within the allowable range of the structure for the application of memory chip 3D-stacked packaging; and (4) in this study, the best warpage value for a memory chip 3D-stacked packaging sample was obtained with an EMC thickness of 1.10 mm, a DAF thickness of 0.01mm, a DAF CTE of $16 \times 10^{-6}/°C$ and a die thickness of 0.13 mm.

Author Contributions: Data curation, S.C. and G.L.; Writing—original draft, X.W.; Writing—review & editing, D.Y. All authors have read and agreed to the published version of the manuscript.

Funding: This research was supported by the Innovation-Driven Development Project of Guangxi Province (Grant No. GuiKeAA21077015), the Key R & D Plan Project of Guangxi Province (grant No. GuiKe AB20159038), the National Natural Science Foundation of China (grant No. 61865004), and the Guilin Science Research and Technology Development Program (grant No. 2020010302).

Institutional Review Board Statement: Not applicable.

Informed Consent Statement: Not applicable.

Data Availability Statement: Not applicable.

Conflicts of Interest: The authors declare no conflict of interest.

References

1. Tsai, M.Y.; Wang, C.T.; Hsu, C.H. The Effect of Epoxy Molding Compound on Thermal/Residual Deformations and Stresses in IC Packages During Manufacturing Process. *Trans. Compon. Packag. Technol.* **2006**, *29*, 625–635. [CrossRef]
2. Tsurumi, N.; Tsuji, Y.; Masago, N.; Yoshizawa, K. Elucidation of Adhesive Interaction between the Epoxy Molding Compound and Cu Lead Frames. *ACS Omega* **2021**, *6*, 34173–34184. [CrossRef] [PubMed]
3. Sumiya, Y.; Tsuji, Y.; Yoshizawa, K. Peel Adhesion Strength between Epoxy Resin and Hydrated Silica Surfaces: A Density Functional Theory Study. *ACS Omega* **2022**, *7*, 17393–17400. [CrossRef] [PubMed]
4. Lin, Y.J.; Hwang, S.J.; Lee, H.H.; Huang, D.Y. Modeling of Viscoelastic Behavior of an Epoxy Molding Compound During and After Curing. *Trans. Compon. Packag. Manuf. Technol.* **2011**, *1*, 1755–1760. [CrossRef]
5. Chao, R.; Fei, Q.; Wang, X. A Finite Element Simulation of PoP Assembly Processes. *J. Beijing Univ. Technol.* **2010**, *38*, 11556068.
6. Cho, K.; Jeon, I. Numerical analysis of the warpage problem in TSOP. *Microelectron. Reliab.* **2004**, *44*, 621–626. [CrossRef]
7. Kim, Y.K.; Park, I.S.; Choi, J. Warpage mechanism analyses of strip panel type PBGA chip packaging. *Microelectron. Reliab.* **2010**, *50*, 398–406. [CrossRef]
8. Srikanth, N. Warpage analysis of epoxy molded packages using viscoelastic based model. *J. Mater. Sci.* **2006**, *41*, 3773–3780. [CrossRef]
9. Kaija, K.; Pekkanen, V.; Maentysalo, M.; Mansikkamaeki, P. Controlling warpage of molded package for inkjet manufacturing. *Microelectron. Eng.* **2008**, *85*, 518–526. [CrossRef]
10. Wu, R.W.; Chen, C.K.; Tsao, L.C. Prediction of package warpage combined experimental and simulation for four maps substrate. In Proceedings of the International Conference on Electronic Packaging Technology & High Density Packaging, Xi'an, China, 16–19 August 2010.
11. Sadeghinia, M.; Jansen, K.; Ernst, L.J. Characterization and modeling the thermo-mechanical cure-dependent properties of epoxy molding compound. *Int. J. Adhes. Adhes.* **2012**, *32*, 82–88. [CrossRef]

12. Ozturk, B.; Gromala, P.; Otto, C.; Fischer, A.; Ernst, L.J. Characterization of Adhesives and Interface Strength for Automotive Applications. In Proceedings of the International Conference on Thermal, Mechanical and Multi-Physics Simulation and Experiments in Microelectronics and Microsystems, Cascais, Portugal, 16–18 April 2012.
13. Kiasat, M.S. Curing Shrinkage and Residual Stress in Viscoelastic Thermosetting Resins and Composites. Ph.D. Thesis, The Delft University of Technology, Delft, The Netherlands, 2000.
14. Scott, T.F.; Cook, W.D.; Forsythe, J.S. Effect of the degree of cure on the viscoelastic properties of vinyl ester resins. *Eur. Polym. J.* **2008**, *44*, 3200–3212. [CrossRef]
15. Driel, W.; Zhang, G.Q.; Janssen, J.; Ernst, L.J.; Su, F.; Chian, K.S.; Yi, S. Prediction and verification of process induced warpage of electronic packages. *Microelectron. Reliab.* **2003**, *43*, 765–774. [CrossRef]
16. Williams, M.L.; Landel, R.F.; Ferry, J.D. The temperature dependence of relaxation mechanisms in amorphous polymers and other glass-forming liquids. *J. Am. Chem. Soc.* **1955**, *77*, 3701–3707. [CrossRef]
17. Mess, L.E.; Brooks, J.M.; Corisis, D.J. Stacked Mass Storage Flash Memory Package. U.S. Patent 7375419, 2008.
18. Duan, Y. Package Structure Design and Process Development for MOSFET Chip Based on Molded Interconnect System Technology. Master's Thesis, Guilin University of Electronic Technology, Guilin, China, 2021.
19. Lin, T.; Chen, C.; Li, J.; Cheng, X.; Qian, W.; Jian, C. Simulation and Validation of FBGA Package Warpage by Considering Viscoelastic Effect. *Semicond. Technol.* **2015**, *40*, 142–147.
20. Jansen, K.M.B. Thermomechanical modeling and characterization of Polymers. Ph.D. Thesis, Delft University of Technology, Delft, The Netherlands, 2005.
21. Ma, D.; He, P.; Xu, Z.; Zhou, Y. *Structure and Properties of Polymers*, 2nd ed.; Science Press: Beijing, China, 1995; pp. 253–267.
22. Yen, F.; Chen, E.; Lai, J.Y.; Wang, Y.P. The Improvement Guidelines of BGA Warpage within SMT Temperature Profile Using Shadow Moiré Technique. In Proceedings of the 8th International Conference on Electronic Packaging Technology, Shanghai, China, 14–17 August 2007.
23. Song, S.N.; Tan, H.H.; Ong, P.L. Die attach film application in multi die stack package. In Proceedings of the Electronic Packaging Technology Conference, Singapore, 7–9 December 2005.
24. Yauw, O.; Jie, W.; Tan, A.; Qin, I.; Schulze, G. Leading edge die stacking and wire bonding technologies for advanced 3D memory packages. In Proceedings of the 19th Electronics Packaging Technology Conference, Singapore, 6–9 December 2017.

coatings

MDPI

Article

Effects of Pd Alloying and Coating on the Galvanic Corrosion between Cu Wire and Bond Pads for a Semiconductor Packaging

Young-Ran Yoo [1] and Young-Sik Kim [1,2,*]

[1] Materials Research Centre for Energy and Clean Technology, Andong National University, 1375 Gyeongdong-ro, Andong 36729, Republic of Korea; yryoo@anu.ac.kr

[2] School of Materials Science and Engineering, Andong National University, 1375 Gyeongdong-ro, Andong 36729, Republic of Korea

* Correspondence: yikim@anu.ac.kr; Tel.: +82-54-820-5504

Abstract: Semiconductor chips are packaged in a process that involves creating a path to allow for signals to be exchanged with the outside world and ultimately achieving a form to protect against various external environmental conditions such as heat and moisture. The wire bonding type of packaging is a method in which thin metal wires are bonded to pads to create an electrical connection between the chip and the lead frame. An Epoxy Molding Compound (EMC) can be applied to protect semiconductor chips from external environmental conditions such as heat, shock, and moisture. However, EMC contains halogen elements and sulfides and has hydrophilic properties, which can lead to a corrosive environment. The present study aims to evaluate the influence of chloride, which is a contaminant formed during the PCB manufacturing process. To this end, the galvanic corrosion of bonding wire materials Cu wire, Cu wire alloyed with 1% Pd, and Cu wire coated with Pd was investigated. The first ball bond was bonded to the Al pad and the second stitch bond was bonded to the Au pad of the manufacturing process, after which the galvanic corrosion behavior in the semiconductor packaging module specimen was analyzed. A model of galvanic corrosion behavior was also proposed.

Keywords: wire bonding; Al or Au pad; Pd alloyed wire; Pd coated wire; galvanic corrosion

Citation: Yoo, Y.-R.; Kim, Y.-S. Effects of Pd Alloying and Coating on the Galvanic Corrosion between Cu Wire and Bond Pads for a Semiconductor Packaging. *Coatings* **2024**, *14*, 544. https://doi.org/10.3390/coatings14050544

Academic Editor: Xu Long

Received: 9 April 2024
Revised: 24 April 2024
Accepted: 26 April 2024
Published: 27 April 2024

1. Introduction

Semiconductor chips are packaged, which is a process that is intended to create a path by which signals can be exchanged with the outside world and create a form that can protect against various external environmental conditions such as heat and moisture [1]. There are several ways to electrically connect semiconductor chips and substrates, including wire bonding, flip chip bonding, and via silicon [2–4]. Among them, wire bonding is a method in which thin metal wires are bonded to pads to create an electrical connection between the chip and the lead frame [5,6]. The metallic types of wires that are mainly used for wire bonding are Au, Ag, Cu, Al, etc., and are typically of fine thickness, which directly affects the efficiency and reliability of semiconductor modules; therefore, it is important to use wires with excellent corrosion resistance and joint degradation characteristics.

In semiconductor packaging, various types of wires are bonded to an Al bond pad to connect them electrically, and then the semiconductor chip is encapsulated to protect it from the external environment. The encapsulation process mainly uses Epoxy Molding Compound (EMC) [7,8]. EMC is widely applied in semiconductor packaging due to its low cost and easy processing, but it inevitably contains ions such as halogen elements (Cl^-, Br^-, F^-) and S [5,9–13], and also has hydrophilic properties that cause it to absorb moisture from the air. As a result, the use of EMC promotes ion diffusion and creates a corrosive environment in the electrolyte [7,12,14–16].

Chloride ions are uniformly distributed in the molding compound matrix. To reduce susceptibility to this corrosion, the mold compound contains ion-trapping components that can immobilize the migration of chloride ions; however, packaging processes such as wire bonding and molding can generate chloride ions in localized areas where the temperatures are in excess of the temperature at which chloride is formed, thus causing high concentrations of chloride ions to diffuse through the mold compound [17].

Some packaging processes attempt to avoid corrosion and reliability problems by adjusting the pH to the neutral range and specifying a low ppm range, but chloride ions can also be introduced when using this method [17]. In other words, impurities in the atmosphere include chloride ions, with particles originating from various types and sources. PVC (polyvinyl chloride) materials containing such chloride may decompose at low temperatures below 135 °C, and in some cases even lower temperatures, depending on the additive ingredients, ultimately forming chlorine compounds such as HCl [17]. These Cl^- ions can weaken or dissolve the passivation layer in a humid environment, consequently initiating corrosion and narrowing the stable pH range as well as accelerating oxidation, reduction reactions, and the growth of intermetallic compounds, all of which are closely related to the failure mechanism of wire bonds [7,12,18]. In corrosive environments, wires and Al bond pads bonded for electrical connections are subject to galvanic corrosion caused by potential differences [4,7,18–21].

In the wire bonding process, the wire bond consists of a first bond, which is made on the die bonding pad, and a second bond, which is made on the substrate bonding pad. The first bond is generally formed into a ball shape to provide adhesion and is called a ball bond, while the second bond is called a stitch bond [22]. Both bonds are formed on the PCB substrate, and because stitch bonds are smaller than ball bonds, they are more susceptible to issues related to the plating quality of the pad and surface contamination, which can affect adhesion [22]. The corrosion of Al pads depends on various factors such as halide concentration, pH, corrosion cell characteristics, bimetal contact, temperature, humidity, etc. [12,23–26]. In Cl^- environments in particular, the potential difference of alloys connected in a galvanic couple in the same corrosion environment influences galvanic corrosion behavior, and the corrosion behavior in a corrosion environment in the same galvanic couple was found to be influenced by the anodic and cathodic Tafel constants of the alloys [27]. It has also been reported that the tendency of galvanic corrosion in PCB units is related to the polarization behavior of the anode material in contact with the cathode material [27].

Cu wire has been widely used as an alternative to Au wire in nano-electronics packaging due to its relatively lower cost of raw materials, which in turn can lower overall costs, as well as advantages such as high electrical conductivity and tensile strength, low thermal and electrical resistivity, and improved long-term reliability compared to Au wire bonding, which are attributed to the low reaction rate between Cu and Al [1,6,9–12,18,28–32]. However, Cu wire requires different process conditions than Au wire, which increases the process cost, and its high hardness makes it difficult to form balls, at the same time requiring the use of high pressures during bonding, which may cause cracks in the pad [6,29]. The Cu wire is also easily oxidized by air due to its low corrosion resistance, so Pd and Au, which are materials with good corrosion resistance, are alloyed or coated on the wire to increase its lifetime and improve the bonding reliability under humid and electrically biased conditions [1,2,5,6,10,29,32–34].

In most cases, wire manufacturers use Cu coated with Pd rather than bare Cu. When Cu wire is coated with Pd, it is resistant to oxidation and does not require a forming gas, so it exhibits a greater reliability of secondary bonding [6]. However, its disadvantage is its higher average hardness and melting point than bare Cu, as well as being more expensive [6]. Pd or Au have very good corrosion properties on their own, but when connected to Al or Cu, the risk of galvanic corrosion increases significantly due to the larger potential difference [9,11,16,35].

The Pd coating on Cu wire shows sufficient potential to replace Au wire, as it has excellent bendability and reliability at a relatively low cost [2,6,34,36,37]. The Pd layer on the Cu wire prevents the formation of CuO and protects the bonded ball from corrosion, and stringent molding compound requirements can also be relaxed [6]. The Pd layer on the Cu wire is beneficial to the Cu wire, reducing its corrosion rate due to its electrochemically noble properties, and it improves the failure time in highly accelerated temperature and humidity stress tests (HASTs) [2,33,38].

As reviewed above, various studies have been conducted to examine corrosion phenomena between wire bonding materials that are typically applied in semiconductor packaging, but a systematic approach to managing galvanic corrosion between these materials is lacking. Therefore, the present study evaluates the effects of chloride contaminants formed during the PCB manufacturing process by assessing galvanic corrosion between Cu wire, Cu wire alloyed with 1% Pd, and Cu wire coated with Pd. This is also explored by fabricating a first ball bond on the Al pad and a second stitch bond on an Au pad, to verify galvanic corrosion behavior using a temperature humidity test (THT) in semiconductor packaging module specimens.

2. Materials and Methods

2.1. Test Specimen

Two types of specimens were used in this study: specimens for electrochemical galvanic corrosion between bonding wires and pads, and module specimens that were bonded under semiconductor packaging simulation conditions.

For electrochemical galvanic test specimens: The pads where wire bonding takes place were created by depositing aluminum (Al) onto a 4-inch silicon (Si) wafer using a DC Magnetron Sputterer (KVS-2002L, Korea Vacuum Tech, Kimpo, Korea). For this purpose, a 99.999% pure Al target was used, with deposition conducted under an argon (Ar) gas flow of 20 Standard Cubic Centimeters per Minute (SCCM), a pressure of 1 mTorr, and a power setting of 50 W to achieve a thickness of 605 nm. Then, the sputtered wafer was diced and connected to copper wire using carbon tape for electrochemical testing. All parts of the electrically connected specimen, aside from an area of 0.04 cm^2, were insulated using acid-resistant epoxy for the electrochemical tests. Meanwhile, the bonding wire was produced from refined materials with increased purity by excluding impurities, melting, and then alloying to determine the wire type. The alloyed wire is initially formed through continuous casting, processed to reduce its diameter, and finally ready for use after undergoing heat treatment and coiling processes. The bonding wires (Cu, Cu1Pd, and PC wire) used in this study were manufactured by MK Electron Co., Ltd. (Yongin, Gyeonggi-do, Korea), and three wires having the composition presented in Table 1 were used as specimens.

Table 1. Types, compositions, and diameters of bonding wires.

Wire Types		Composition	Diameter, μm
Pure Cu	Cu wire	Cu 99.99 wt. %	25
Pd alloying	Cu1Pd wire	Cu 99 wt. % + Pd 1 wt. %	25
Pd coated Cu	PC wire	Cu 99.99 wt. %/Pd coating	25

For the temperature humidity test wire bonding module [27]: The wire bonding module consists of a first ball bond and a second stitch bond. Under the conditions listed in Table 2, wire bonding was performed on the bond pads in the FR4 PCB using a wire bonder (K&S RAPID PRO, Kulicke & Soffa, Singapore, Singapore), and the modules were manufactured by MK Electron Co., Ltd. (Yongin, Korea). The first ball bond was bonded with a wire of 25 μm in diameter to a pure Al pad, and the thickness of the Al pad was 10,000~12,000Å. The second stitch bond was bonded with a 25 μm diameter wire to a pure Au pad (the thickness of the Au pad is 0.3 μm). Figure 1 shows a module specimen manufactured as detailed above.

Table 2. Bond parameters of wire bonding on the module.

	Cu Wire			Cu1Pd Wire			PC Wire		
Process	**1st**	**2nd**	**EFO ***	**1st**	**2nd**	**EFO ***	**1st**	**2nd**	**EFO ***
Current (mA)	55~85	75~120	40	55~85	75~120	40	55~85	75~120	40
Time (μs)	5~10	10	730	5~10	10	710	5~10	10	705
Force (grams)	25~70	45~75		25~70	45~75		25~70	45~75	

* EFO: Electronic Flame-Off.

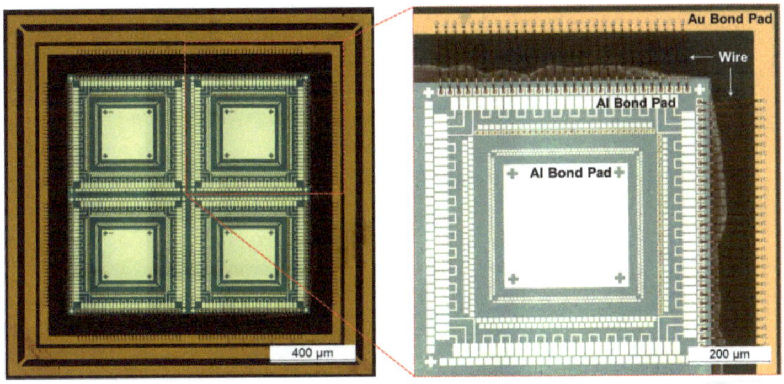

Figure 1. Module specimen containing 1st ball bond and 2nd stitch bond manufactured by MK electron Co., Ltd.

2.2. Polarization Test

A polarization test was conducted to evaluate the corrosion characteristics of the bond pad and bonding wire. To evaluate changes in this behavior under different hydrochloric acid (HCl) concentrations, solutions of 1% NaCl, 1% NaCl + 0.01% HCl, 1% NaCl + 0.1% HCl, and 1% NaCl + 1% HCl were de-aerated with N_2 gas at a rate of 200 mL/min for 30 min before the polarization test. The solution temperature was kept at 25 °C. The test equipment used was a Potentiostat (Interface 1000, Gamry, Warminster, PA, USA), in which the reference electrode was a saturated calomel electrode (SCE) and the counter electrode was a platinum electrode. After installing the specimen in the Avesta cell, the corrosion potential was measured after 60 s of conditioning at a potential approximately 200 mV below the corrosion potential, and with an initial delay of 150 s at open-circuit potential, after which the polarization test was performed at a rate of 0.33 mV/s.

2.3. Electrochemical Galvanic Corrosion Test

An electrochemical galvanic corrosion test was conducted to evaluate the galvanic corrosion behavior of the bond pad and bonding wire. In this test, an SCE was used as the reference electrode and a bonding wire was used as the counter electrode. The test solution used was the same 1% NaCl + x% HCl solution used in the polarization test, with the anode to cathode ratio set to 1:1 to exclude any surface. The test equipment used was a Potentiostat (Interface 1000, Gamry, Warminster, PA, USA).

2.4. Temperature Humidity Test (THT)

To evaluate the corrosion properties of the module specimens, a temperature humidity test was performed at a constant temperature and using a humidity chamber (PR-2J, ESPEC, Osaka, Japan). For wire-bonded modules, the test was performed after they were baked at 175 °C for 3 h prior to the test for IMC (intermetallic compound) growth. The module specimen was placed in a chamber by fixing the module specimen to an epoxy substrate and placing it in a glass reaction chamber containing the test solution (the specimen assembly schematic that was followed is described in detail in a previous work [28]). The specimens were placed in the glass cell, which was then placed in the chamber, and the chamber was maintained at 85 °C, 85% R.H. for 100 h. The test solutions were 1% NaCl solution and 1% NaCl + 0.1% HCl solution.

2.5. Surface Analysis

An Optical Microscope (ZEISS, Axiotech 100HD, Oberkochen, Germany) was used to observe the specimen morphology before and after testing, while a 3D microscope (VK-X3000, Keyence, Itasca, IL, USA) was used to observe the overall geometry of the wire connected to the 1st ball bond and the 2nd stitch bond before and after the test. Meanwhile, FE-SEM (FE-SEM, MIRA3XMH, Tescan, Brno, Czech Republic) and EDS (Energy Dispersive X-ray Spectrometer, Mmax 50, Oxford, UK) were used to observe and analyze the morphology of the corroded surfaces and components, respectively.

3. Results and Discussion

3.1. Electrochemical Galvanic Corrosion Behavior of Bonding Wires and Pad Alloys

To evaluate the galvanic corrosion behavior at different HCl concentration, test solutions were prepared by increasing the HCl concentration in a 1% NaCl solution to 0.01%, 0.1%, and 1%. The polarization behavior was then measured in a test solution prepared for every single specimen, and the corrosion rates obtained by Tafel analysis are displayed in Figure 2. Figure 2 shows the effect of HCl concentration on the corrosion rate of sputtered Al and bonding wires as a single specimen in de-aerated 1% NaCl + x% HCl solution at 25 °C. For the Al specimen for the first bond pad, with the increase in HCl concentration, the corrosion rate of sputtered Al tended to decrease and then increase again, but the corrosion rate was very low in the order of μm/y, indicating that the effect of HCl was not significant. According to a previous work by the research team that authored this work [27], it was confirmed that this low corrosion rate can be attributed to the difference in the crystal structure, in which part of the Al becomes amorphous. Meanwhile, the second bond pad materials, Au, Cu wire, Cu1Pd alloy wire, and Pd coated Cu wire, showed a tendency whereby the corrosion rate increased depending on the HCl concentration, but this rate still showed a very small value compared to that observed when using Al.

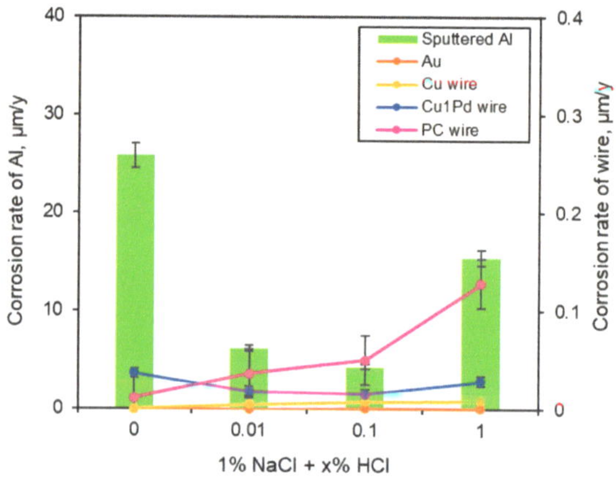

Figure 2. Effect of HCl concentration on the corrosion rate of sputtered Al and bonding wires as a single specimen in de-aerated 1% NaCl + x% HCl solution at 25 °C.

First, to investigate the galvanic corrosion behavior between the sputtered Al and the bonding wire for the first bonding, the effects of Pd alloying and Pd coating on the galvanic corrosion behavior of the Cu-based wire and the Al bond pad when connected in a galvanic couple were examined. The test solution was a 1% NaCl solution with increased HCl concentrations of 0.01%, 0.1%, and 1%. The galvanic corrosion current density and galvanic corrosion potential were measured by connecting the sputtered Al to the working electrode and connecting each wire to the counter electrode. Figure 3 shows the effect of HCl concentration on the electrochemical galvanic corrosion between the bonding wires and the Al bond pad. Figure 3a,b show the galvanic corrosion potential and current density between the Cu wire and the sputtered Al. Increasing the HCl concentration in the 1% NaCl solution increased the galvanic corrosion potential, but did not appear to have a marked effect on the galvanic corrosion current density. Figure 3c,d shows the galvanic corrosion potential and current density between the Cu1Pd wire and the sputtered Al. The results of the galvanic corrosion behavior measured for 1800 s show that the galvanic corrosion current density increased with increasing concentration of HCl. However, no clear concentration effect could be clearly observed, and it was also observed that the galvanic corrosion potential had a slight concentration effect. Figure 3e,f show the galvanic corrosion potential and current density between the PC wire and the sputtered Al. The change in galvanic corrosion potential with increasing HCl concentration was minor, while the galvanic corrosion current density showed an increasing trend with increasing HCl concentration.

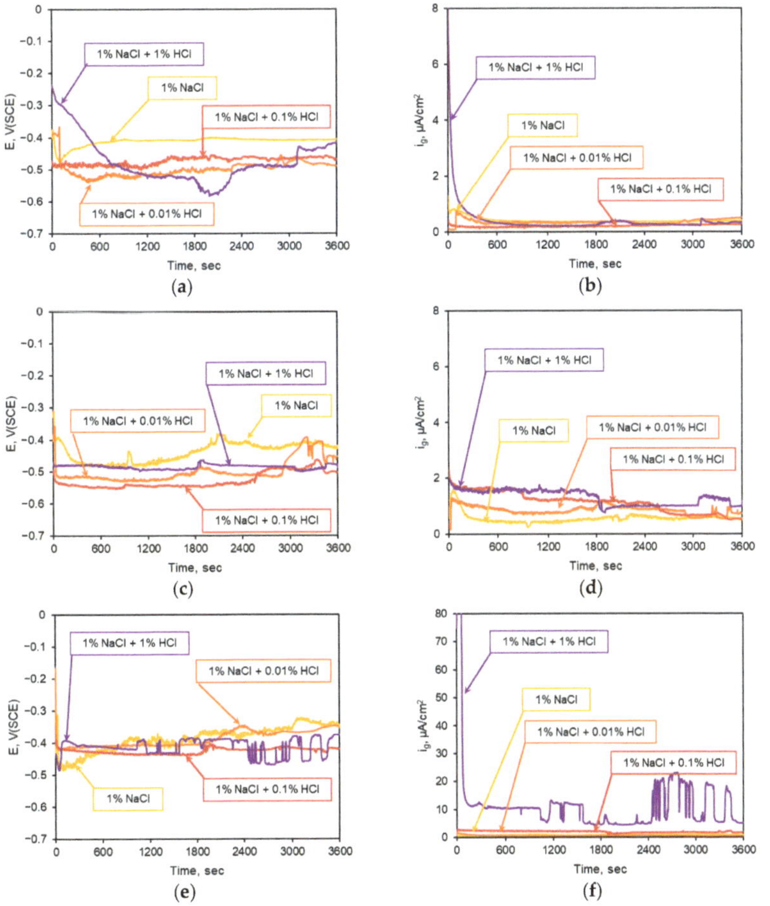

Figure 3. Effect of HCl concentration on the electrochemical galvanic corrosion between the bonding wires and the Al bond pad: (**a**) Galvanic corrosion potential of Cu-Al [27] and (**b**) Galvanic current density of Cu-Al. (**c**) Cu1Pd-Al galvanic corrosion potential and (**d**) Cu1Pd-Al Galvanic current density. (**e**) PC-Al galvanic corrosion potential and (**f**) PC-Al galvanic current density.

The galvanic corrosion rate was calculated based on the galvanic test results depicted in Figure 3, and the resulting corrosion rates are shown in Figure 4. Figure 4 shows the effect of HCl concentration on the galvanic corrosion rate of sputtered Al coupled with Cu wires as measured through electrochemical galvanic corrosion testing. When Cu and Cu1Pd wires and sputtered Al were galvanically coupled, the effect of HCl concentration was small. However, for the galvanic couple of PC wires with sputtered Al, the corrosion rate tended to increase with increasing HCl concentration. This phenomenon is attributed to the fact that the galvanic corrosion rate is further increased by the large potential difference, because the Pd coated on Cu is more noble than Cu itself [1].

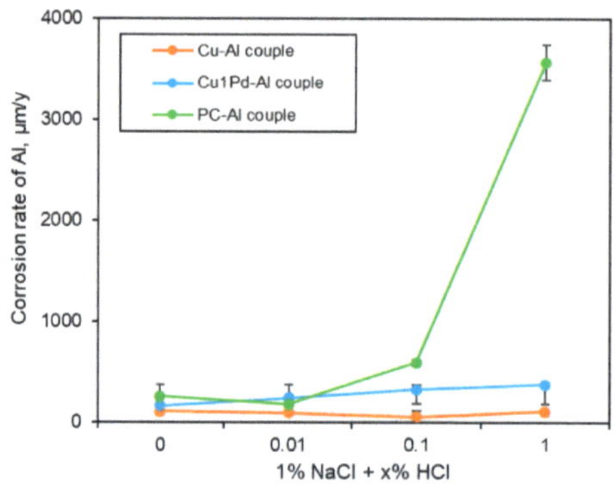

Figure 4. Effect of HCl concentration on the galvanic corrosion rate of sputtered Al coupled with Cu wires, as measured by the electrochemical galvanic corrosion test.

Figure 5 shows the surface appearance, documented by optical microscopy (OM), of Al after the electrochemical galvanic corrosion test between sputtered Al and Cu, Cu1Pd and PC wire. It can be seen from the figure that the corrosion of Al accelerated with increasing HCl concentrations. The results show similar trends not only for Cu wire, but also for Pd alloying and coating.

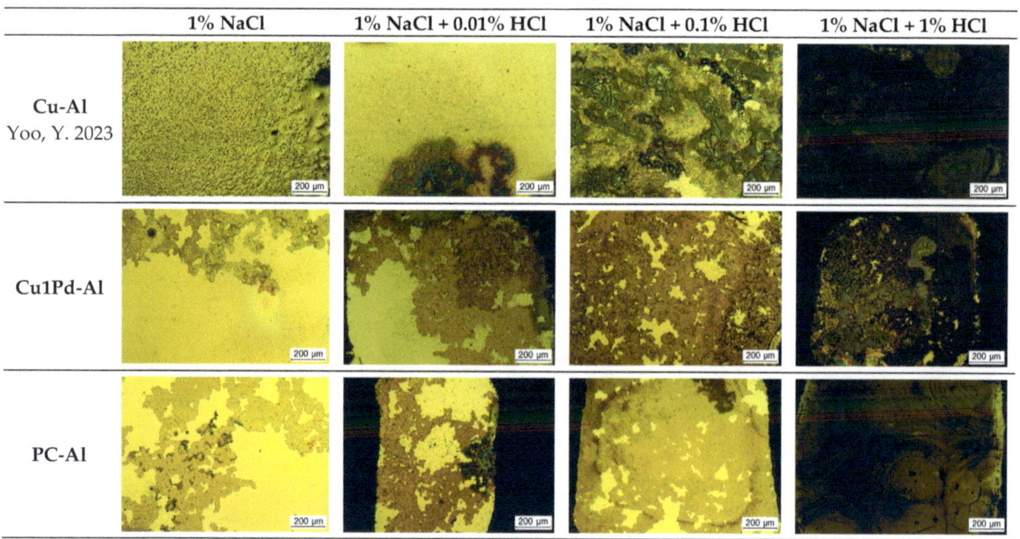

Figure 5. Effect of HCl concentration on the surface appearance of Al coupled with Cu wires (OM) [27].

The morphology of the corroded Al surface, where corrosion was accelerated with an increasing HCl concentration, was observed using SEM, and the surface components were analyzed by EDS. Figure 6 shows the effects of the HCl concentration on the surface appearance of Al coupled with Cu wires (SEM) and on the elemental distribution of sputtered Al coupled with Cu wires after the electrochemical galvanic corrosion test. In the case of the Cu wire, as the HCl concentration increased, the Al surface became more and more covered with corrosion products and more cracks were formed. At 1% HCl, the Al corrosion products in the crack area were locally delaminated and the substrate metal was exposed. The results of the EDS surface analysis confirm the course of the corrosion process according to the HCl concentration. Up to the concentration of 1% NaCl + 0.1% HCl, the content of Al increased due to the formation of Al oxide, a corrosion product, on the surface of Al. However, as the concentration of HCl increased, cracks and the delamination of Al oxide occurred, which ultimately exposed the base metal, Si; as a result of this process, the concentration of Si increased and the concentration of Al decreased. In the case of Cu1Pd wire, the surface image obtained by SEM and the average value from the surface analysis show that, as the concentration of HCl increased, the corrosion products formed on the surface increased, and the content of Al decreased due to the dissolution of Al, as a result of which Si, a base metal, was exposed. For the PC wire and sputtered Al connected via galvanic coupling, the corrosion morphology of Al as a function of HCl concentration was analyzed by SEM-EDS, and the results show that the dissolution of Al accelerated with increasing HCl concentrations. It can be seen that, at a concentration of 1% HCl, all the Al was dissolved, and Si and Si oxide were formed on the surface.

To analyze the galvanic corrosion behavior detailed above, an electrochemical analysis was conducted using mixed potential theory [1]. Figure 7 shows the analysis of galvanic corrosion using the mixed potential theory between sputtered Al specimen and Cu wires. From the polarization curves of two alloys intersecting in the galvanic couple, various electrochemical parameters can be obtained, including the corrosion potential difference (ΔE), the galvanic corrosion potential (E_g) and the corrosion current density according to the mixing potential theory, the anodic reaction Tafel constant (β_A), and the cathodic reaction Tafel constant (β_C).

The sputtered Al pad, which is connected to three types of Cu wires in the galvanic couple, has a relatively active potential, so it can serve as an anode. As the cathodic polarization curves of the three types of Cu wires meet in the trans-passive region of the anodic polarization curve of sputtered Al, the corrosion of Al becomes accelerated. Moreover, as the concentration of HCl increases, the galvanic current density at which each polarization curve intersects tends to increase, indicating that as the corrosivity of the test solution increases, the galvanic corrosion rate also increases. In the case of the Cu1Pd wire, we aimed to improve the corrosion resistance of Cu by alloying with Pd, a metal with good corrosion resistance, but alloying in a small amount of 1% causes a cathodic polarization behavior similar to that of pure Cu wire. The electrochemical galvanic corrosion behavior of Cu1Pd wire with an Al pad is also similar to that of Cu wire. On the other hand, in the case of the Pd-coated PC wire, the corrosion potential increases slightly in the 1% NaCl solution due to the noble Pd, and the cathodic polarization behavior is similar to that of the Cu wire. However, as the corrosivity of the test solution increases, the coated Pd in the thin film also dissolves, and galvanic corrosion with Cu increases, increasing the current density due to cathodic polarization. For this reason, it can be seen that the corrosion rate increases as a result of the galvanic corrosion of the PC wire and Al pad.

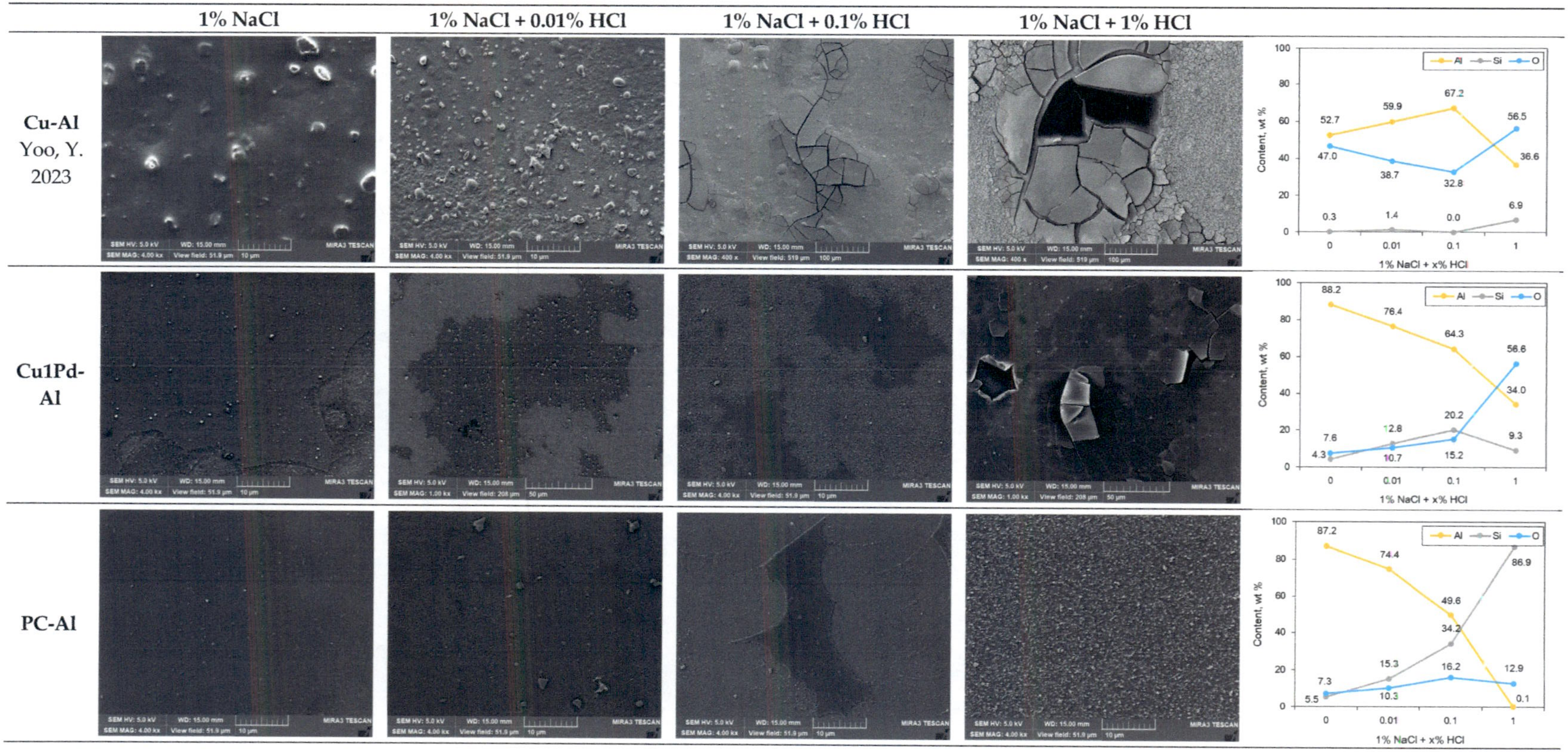

Figure 6. Effect of HCl concentration on the surface appearance of Al coupled with Cu wires (SEM) and on the elemental distribution of Al coupled with Cu wires after electrochemical galvanic corrosion tests [27].

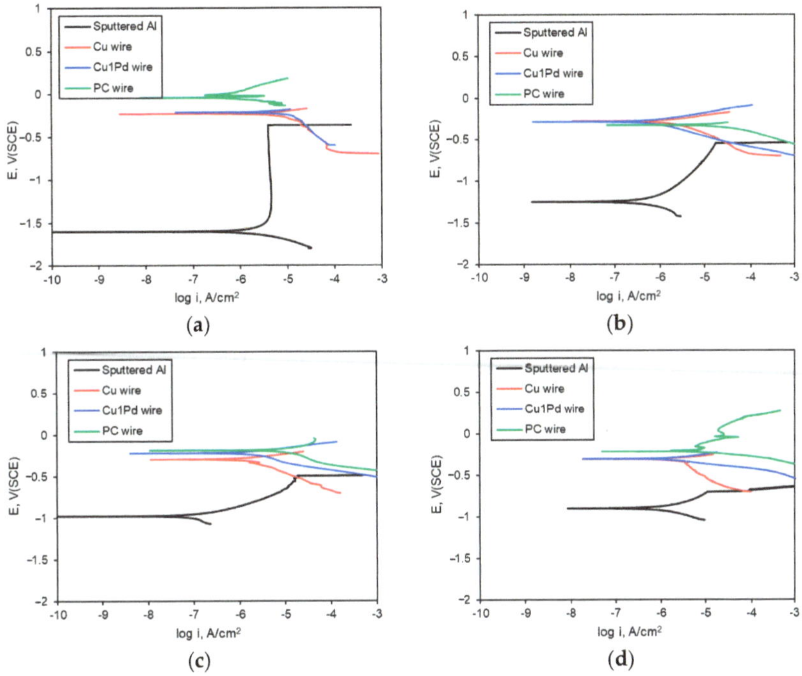

Figure 7. Analysis of galvanic corrosion using mixed potential theory between sputtered Al specimen and Cu wires; (**a**) 1% NaCl, (**b**) 1% NaCl + 0.001% HCl, (**c**) 1% NaCl + 0.01% HCl, (**d**) 1% NaCl + 0.1% HCl.

The galvanic corrosion behavior was analyzed by various electrochemical parameters obtained from the polarization behavior of each alloy, and the results are shown in Figure 8. In particular, Figure 8a,b show the effects of Pd alloying and Pd coating on the galvanic potential difference (ΔE) when the three types of wires and Al bond pads are galvanically connected. In other words, when a Pd alloying or Pd coating was applied to the Cu wire connected to Al in a galvanic couple, the galvanic potential difference increased regardless of the HCl concentration. In general, in the same environment, the galvanic potential difference is highly correlated with the galvanic corrosion rate, since the galvanic corrosion rate increases as the galvanic potential difference increases (Figure 8a).

However, as shown in Figure 8b, the galvanic potential difference decreased with increasing HCl concentration, which shows the difficulty associated with explaining the effect of HCl concentration on the intermetallic galvanic potential difference in the same material. To analyze this behavior, the anode and cathode Tafel constants were compared as a function of HCl concentration. Figure 8c,d shows the anodic Tafel constant and cathodic Tafel constant depending on the HCl concentration. For sputtered Al, the anodic Tafel constant decreased with increasing HCl concentration when the bonding wires and bonding pads were connected via galvanic coupling. The cathodic Tafel constant of the wires was also generally reduced. According to the latest model reporting the effects of HCl concentration on galvanically coupled Au and Cu wires and sputtered Al [27], for the same material, changing the Tafel constant of each galvanic coupled affects the galvanic corrosion rate when the corrosion environment changes. Therefore, in the case of the Cu1Pd-Al galvanic couple in this study, the anodic Tafel constant of Al and the cathodic Tafel constant of the wire decreased with the increase in HCl concentration. Meanwhile, in the case of the PC-Al galvanic couple, the anodic Tafel constant of Al decreased and the cathodic Tafel constants of the wire decreased with increasing HCl concentrations. In other words, as shown in Figure 4, the effects of HCl concentration on the galvanic corrosion of Al pads and three types of wires can be interpreted based on a combined consideration of the anodic and cathodic Tafel constants.

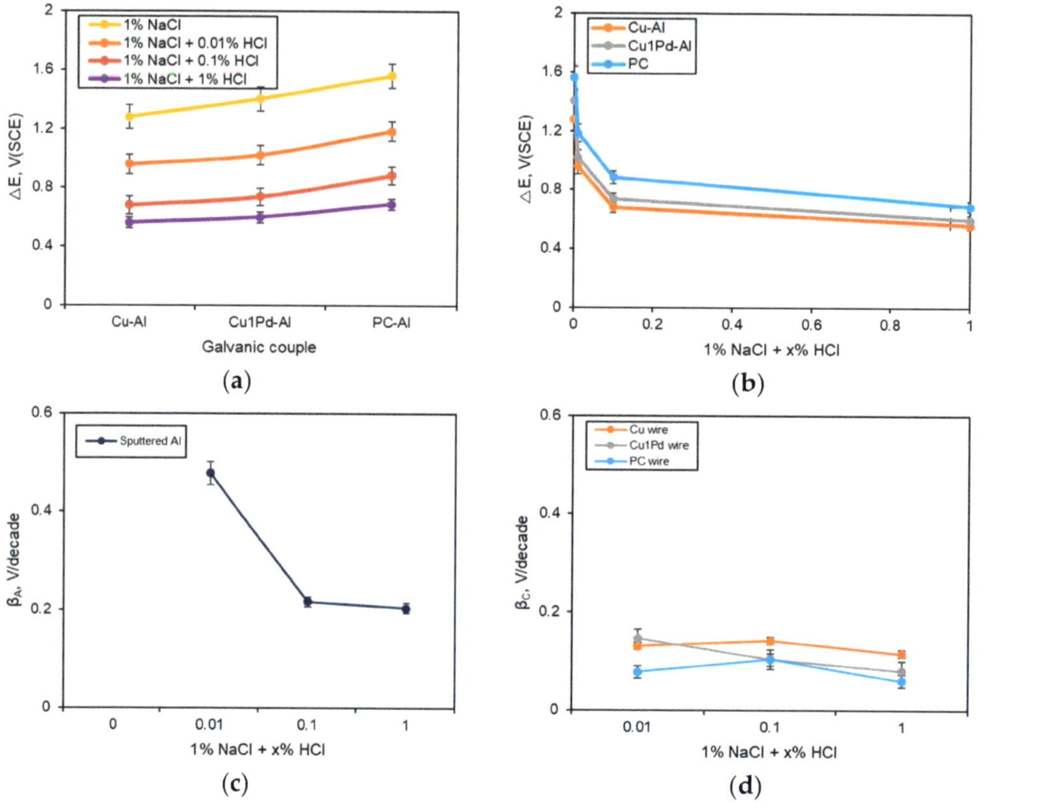

Figure 8. Analysis of galvanic corrosion behavior: (**a**) Effect of Pd and (**b**) effect of HCl concentration on galvanic potential difference, (**c**) anodic Tafel constant and (**d**) cathodic Tafel constant on HCl concentration.

3.2. *Galvanic Corrosion Behavior of the Wire Bonding Module by THT*

Meanwhile, to observe the galvanic corrosion behavior at the interface of the wire bonded to the PCB bonding pad, THT was conducted for 100 h at 85 °C and 85% using a wire bonding module. The test solutions were 1% NaCl and 1% NaCl + 0.1% HCl, and the surface morphology and elemental distribution of the first ball bonding part and second stitch bonding part were analyzed after the test. Figure 9 shows the surface appearance of the wire bonding module before and after THT at 85 °C and 85% relative humidity for 100 h. Before testing, there were no signs of corrosion, and it could be seen that both the first ball bonding, which is made of a square die-shaped pure Al pad, and the second stitch bonding area, which is bonded to the pure Au substrate by pressing, maintained their integrity (Figure 9a). Meanwhile, after THT, it can be seen that the wire, bond pad, and bonding area were severely corroded, regardless of the wire type (Figure 9b–d). In the first ball bonding area, corrosion progress could be seen in the Al bonding pad area, while in the second stitch bonding area with the Au substrate, where most of the Au maintained its integrity, it could be seen that the bonding area and wire were corroded. In particular, as shown in Figure 9d, it can be seen that the PC wire module is covered with corrosion products over a wide area.

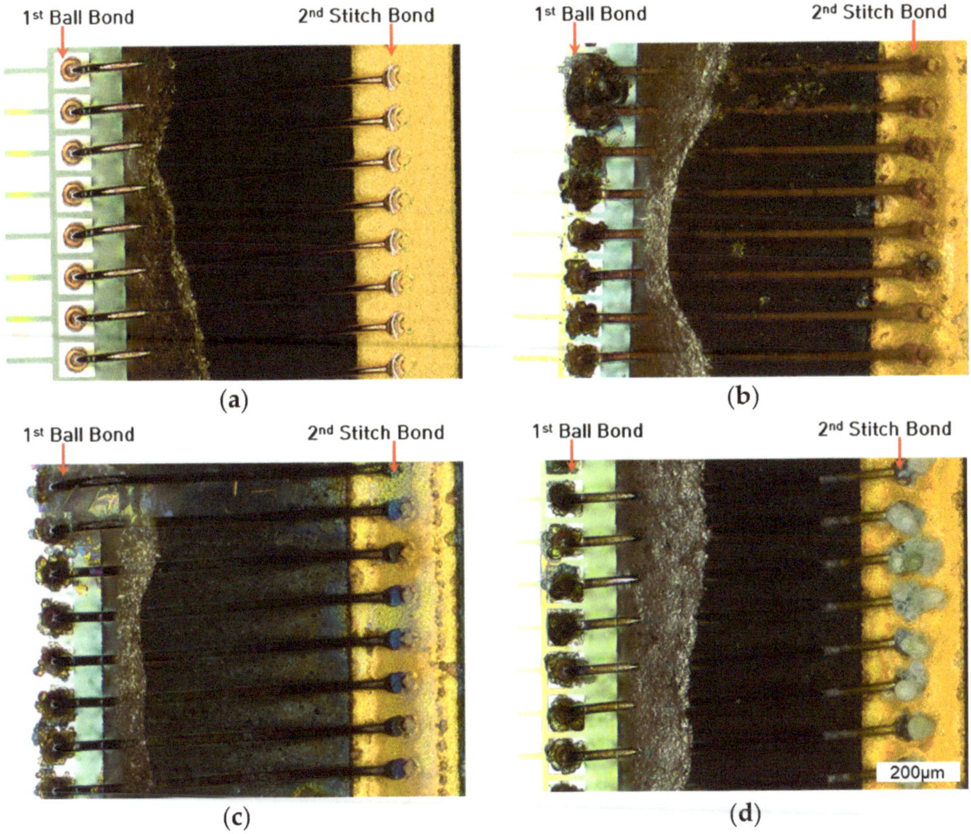

Figure 9. Surface appearance of a wire bonding module after THT at 85 °C and 85% relative humidity for 100 h: (**a**) before test, (**b**) Cu wire module, (**c**) Cu1Pd wire module, (**d**) PC wire module.

For each module, the surface morphology and elemental distribution of the first ball bonding area and the second stitch bonding area were confirmed by SEM-EDS. Figure 10 shows the SEM images and elemental distribution of the first bond area with wires and the Al pad on the module after temperature–humidity tests at 85 °C, 85 % RH in 1% NaCl. In the first ball bond area between Cu wire and Al pad, when a THT was conducted in a 1% NaCl solution for 100 h, it could be seen that the Al pad near the ball was corroded and corrosion products were formed, and it could also be observed that the bonded Cu ball also formed corrosion products. A similar tendency was observed in the first ball bond area of the Cu1Pd wire and the PC wire module. Most corrosion products were formed on the Al pad, and as the amount of corrosion products increased, they were increasingly located in the ball area. These corrosion products were mainly composed of Al oxide, and Al, which has a relatively active potential, is preferentially corroded by the galvanic effect, which is consistent with the electrochemical galvanic corrosion behavior observed herein. Cu wires were also found to be partially corroded due to their low resistance in highly corrosive environments, leading to the formation of Cu oxide corrosion products on the surface.

Figure 11 shows the SEM images and elemental distribution of the first bond area with wires and the Al pad on the module after THT at 85 °C, 85% RH in 1% NaCl + 0.1% HCl. When the Cu wire, Cu1Pd wire, and PC wire were exposed to the test solution for 100 h at the first ball bond between the Cu wire, Cu1Pd wire, PC wire, and Al pad, the Al pad near the ball was corroded and the Al oxide was corroded. The resulting corrosion products became attached to the bonded wire balls. It was also confirmed that, as the severity of the corrosive environment increased, Cu, the main component of the wire, dissolved and formed corrosion products in the form of Cu oxide.

In the case of the PC wire, the Pd coated on the surface can be clearly detected in the component distribution after THT in a 1% NaCl environment (Figure 10, third row). However, in the component distribution results after THT in a 1% NaCl + 0.1% HCl environment (Figure 11, third row), the very thin Pd coating layer is not clearly revealed, and it is more difficult to detect because it is also corroded due to the strong corrosive environment. Ultimately, the results indicate that Cu oxide forms together with Al oxide on the Al pad, ball bond area, and wire surface.

It can be seen that the galvanic corrosion behavior in these packing modules is consistent with the electrochemical galvanic corrosion results. In the wire bonding module, when the Al pad and the first ball bonded wire are exposed to the corrosive environment, galvanic corrosion occurs, which accelerates the corrosion of the relatively active Al pad. However, when a highly corrosive environment is formed, the wire also corrodes, as shown in Figures 3 and 7.

When the Au pad and wire are exposed to a corrosive environment, the corrosion of Au, which is a more noble metal than the wire, is suppressed, and the wire material acts as an anode, thus causing galvanic corrosion that accelerates the corrosion in general. Figure 12 shows the SEM images and elemental distribution of the second stitch bond area with wires and the Al pad on the module before the test. The wires pressed into the Au pad do not have any oxides formed on them, and there are no signs of corrosion. In the case of the Cu1Pd wire, 1% of the Pd was alloyed, so the Pd component was not clearly visible, while in the case of the PC wire, only Cu was present in the compressed area, and the Pd component was detected due to the Pd coating on the surface of the wire.

Figure 13 shows the SEM images and elemental distribution of the second stitch bond area with wires and the Au pad on the module after the temperature–humidity test at 85 °C, 85% RH in 1% NaCl. The surface morphology of the second bond area of the Cu wire and Au pad shows that the corrosion of the Cu wire progresses with the formation of corrosion products covering the surface of the Au pad. In the case of the Cu1Pd wire, no corrosion products were formed under the influence of 1% NaCl solution for 100 h, unlike in the case of the Cu wire. However, it was confirmed that Cu oxide was partially distributed on the surface of the Au pad due to the dissolution of the Cu wire. On the other hand, in the case of the PC wire, it was observed that more corrosion products were formed and covered on the Au pad than in the case of the Cu wire. When the environment became more corrosive after adding HCl, the corrosion of the second stitch bond area was also accelerated.

Figure 14 shows the SEM images and the elemental distribution of the second stitch bond area with wires and the Au pad on the module after the temperature–humidity test at 85 °C, 85% RH in 1% NaCl + 0.1% HCl. In the case of the Cu wire, the formation of corrosion products could be observed, which spread to the surface of the Au pad, thereby covering the Au surface. More corrosion products were formed than in the 1% NaCl environment, and the results also showed a trend whereby, as the intensity of the corrosion environment increased, the amount of corrosion products also increased. The Cu1Pd wire also showed a similar trend to the Cu wire. However, the PC wire showed the most severe corrosion behavior, as occurred in the 1% NaCl environment. This result is attributed to the fact that the Pd covering the surface is a nobler metal than Cu, so the exposed area of Cu acting as an anode via stitch bonding is small, and the area of Au and Pd acting as cathodes increases, ultimately resulting in a larger galvanic effect.

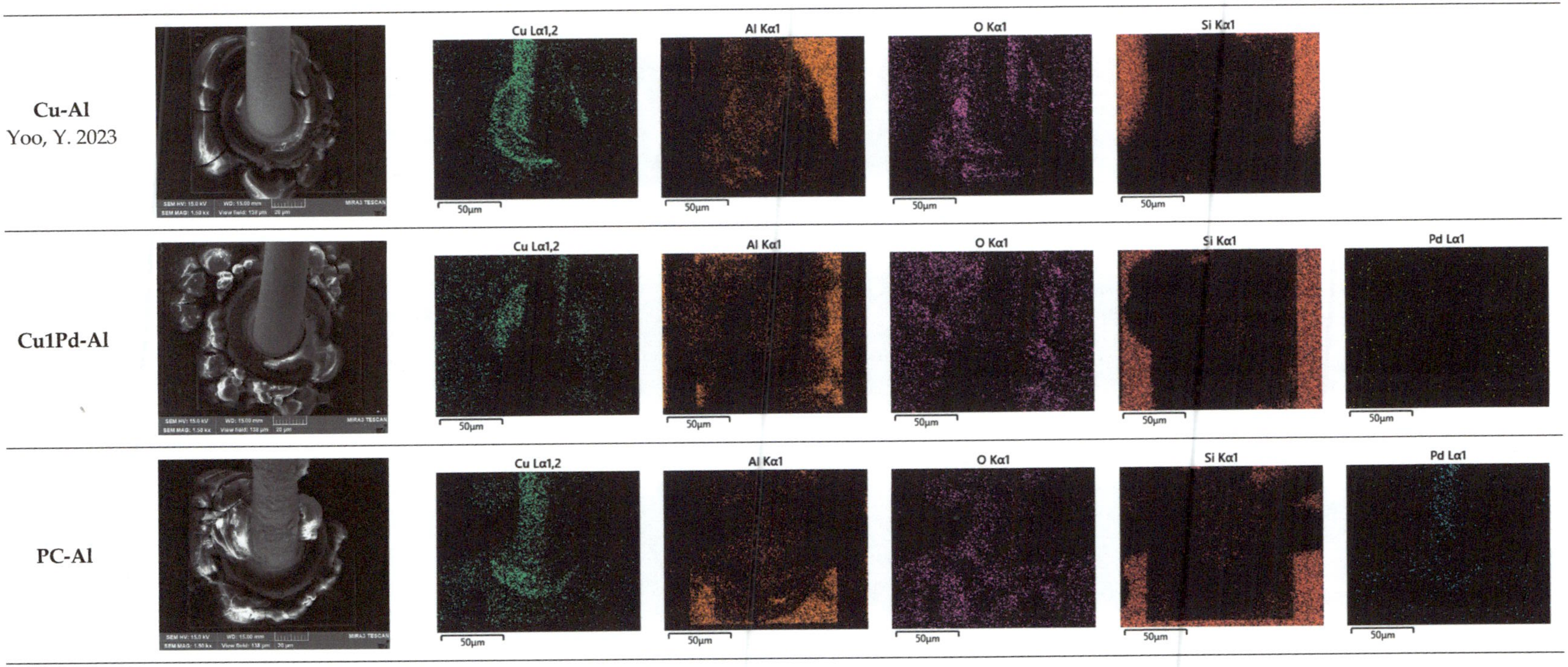

Figure 10. SEM images and elemental distributions of the first ball bond area with wires and the Al pad on the module after the temperature–humidity test at 85 °C, 85% RH in 1% NaCl [27].

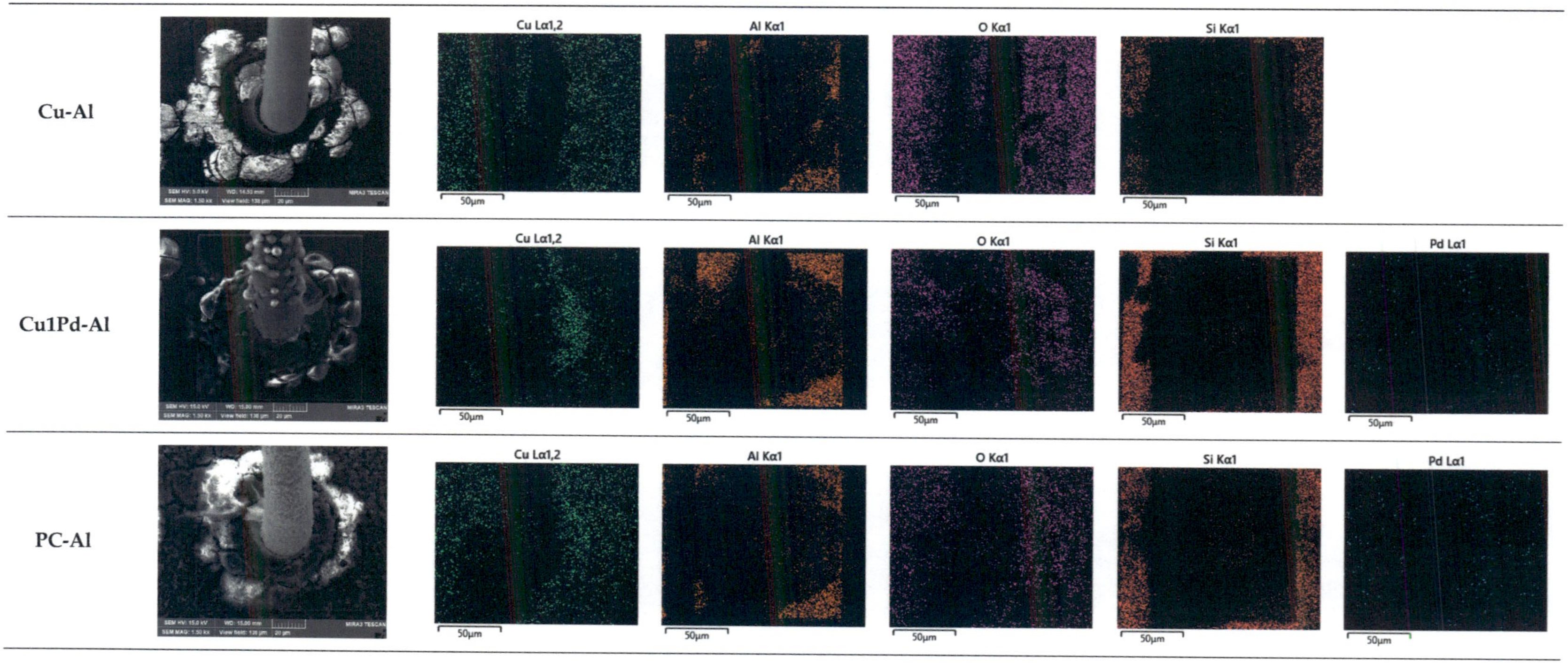

Figure 11. SEM images and elemental distributions of the first ball bond area with wires and the Al pad on the module after the temperature–humidity test at 85 °C, 85% RH in 1% NaCl + 0.1% HCl.

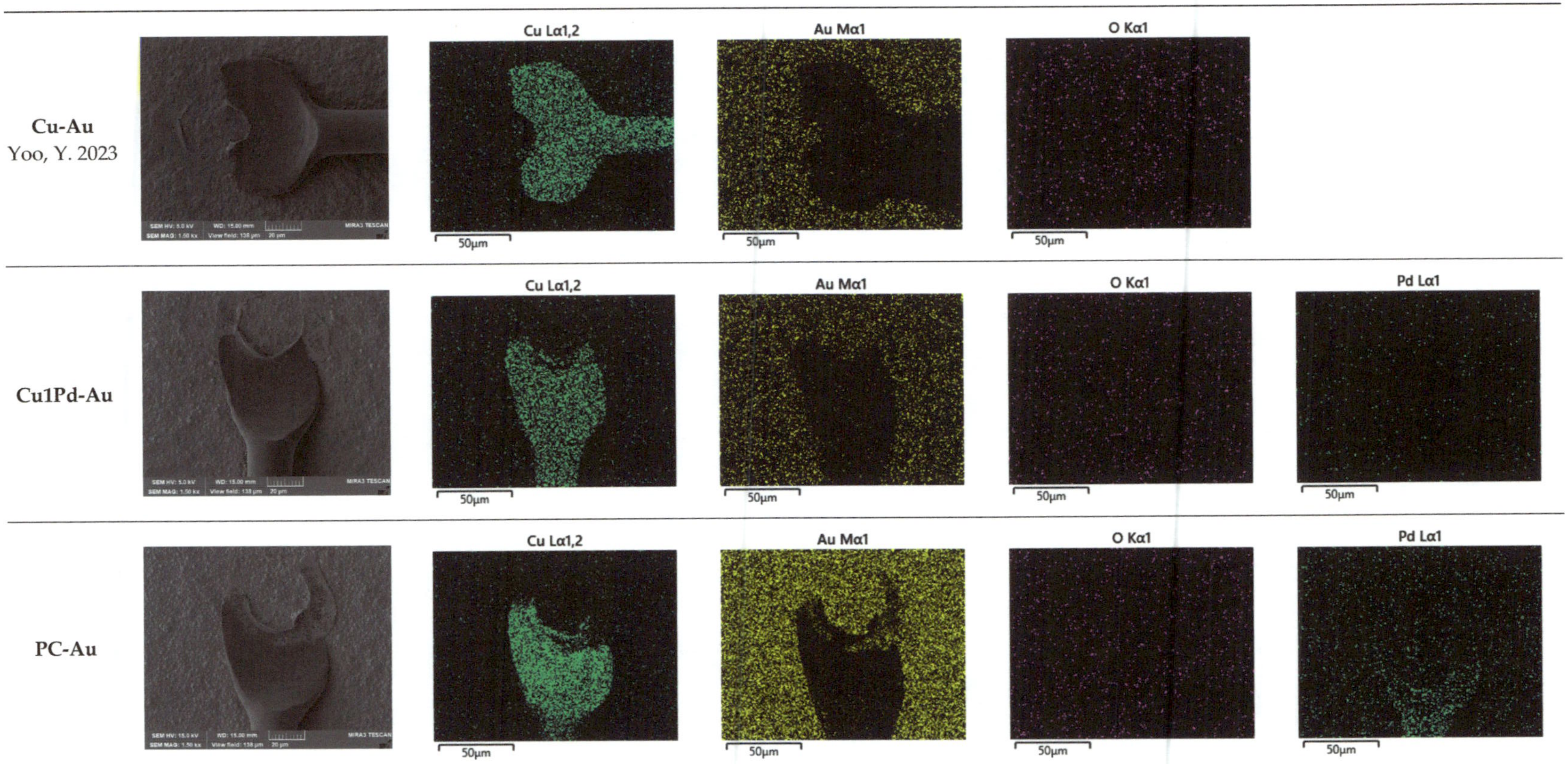

Figure 12. SEM images and elemental distributions of the second stitch bond area with wires and the Au pad on the module before testing [27].

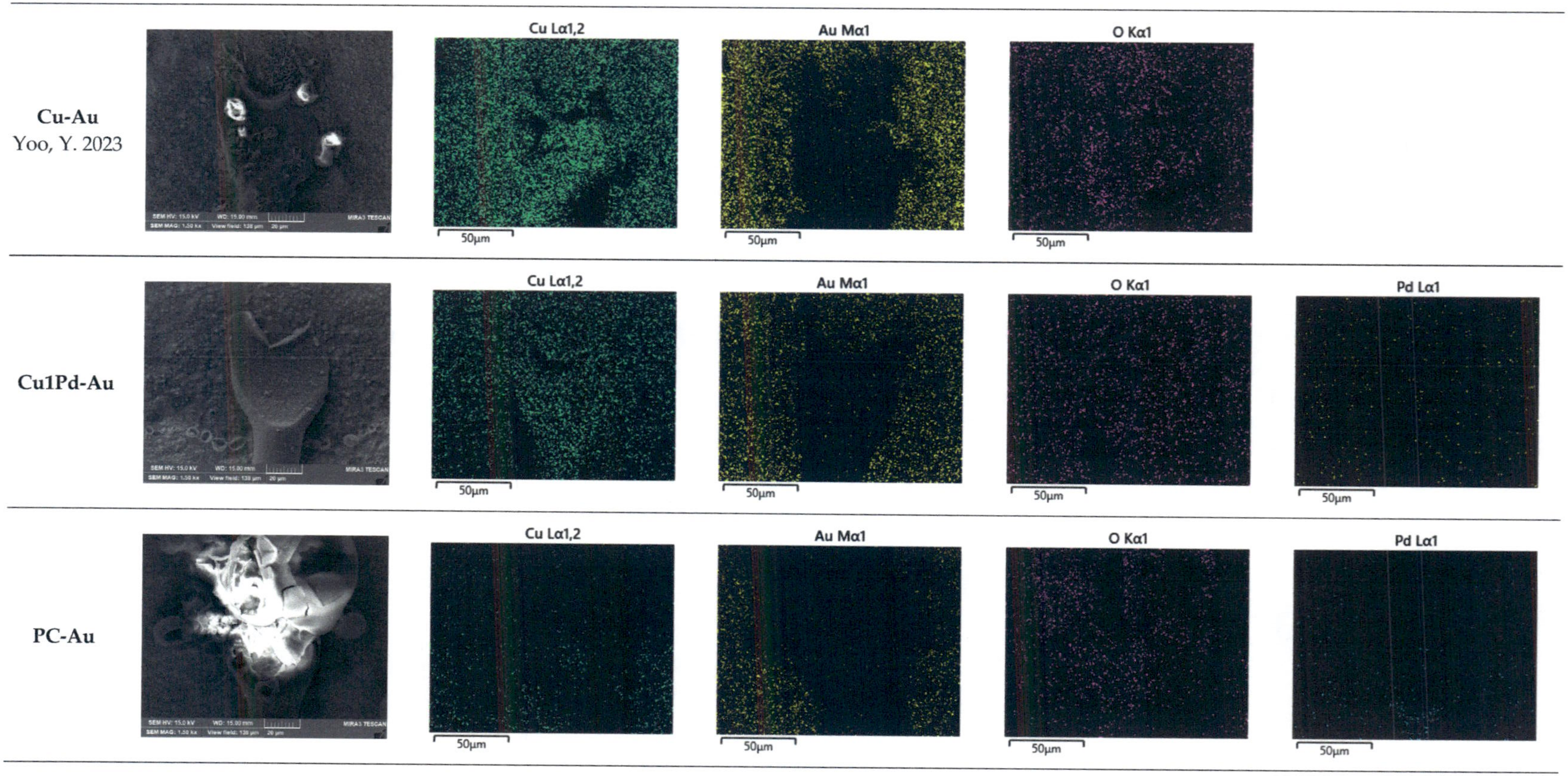

Figure 13. SEM images and elemental distributions of the second stitch bond area with wires and the Au pad on the module after the temperature–humidity test at 85 °C, 85% RH in 1% NaCl [27].

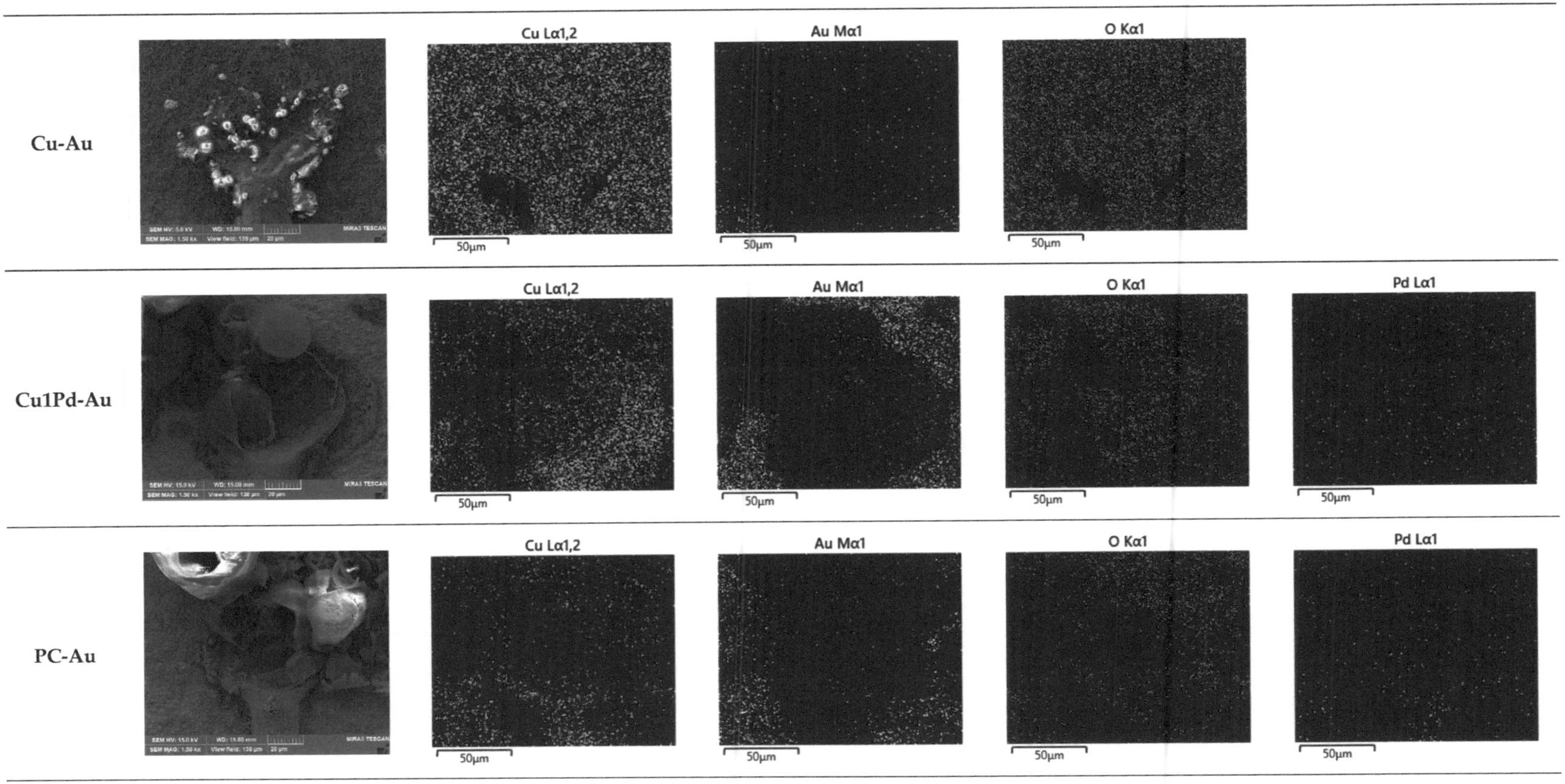

Figure 14. SEM images and elemental distributions of the second stitch bond area with wires and the Au pad on the module after the temperature–humidity test at 85 °C, 85% RH in 1% NaCl + 0.1% HCl.

As detailed above, the galvanic corrosion behavior occurring in the second stitch bonding of the packaging module was analyzed using mixed potential theory. Figure 15 shows the analysis of galvanic corrosion behavior using mixed potential theory between the Au specimen and Cu wires. Figure 15a shows a graph comparing the polarization behavior of Au and Cu wires in a 1% NaCl solution. The corrosion potential of Au is higher than that of other examined wires other than PC wire, which means that, when Au- and Cu-based wires are connected into a galvanic couple, the wire works as an anode, and galvanic corrosion occurs. On the other hand, in the case of Pd-coated PC wire, the wire material exposed to the initial corrosive environment becomes Pd, a noble material, and thus exhibits a corrosion potential similar to Au. However, there is a difference of about two orders of magnitude in the current density, so it can be said that the Cu substrate also affects the corrosion rate. Therefore, for the second stitch bonding area where the PC wires are bonded, the Cu is exposed at the part pressed by bonding, and then the accelerated galvanic corrosion is caused by the Au bond pad and Pd coating layer. As the intensity of the corrosive environment increases, it becomes easier for Cu to be exposed, as it is less protected by the thin Pd layer. As can be seen in Figure 15b, galvanic corrosion is accelerated when the anodic polarization behaviors of all wires interact with the cathodic polarization behavior of Au. Figure 16 show the effect of HCl concentration on the galvanic corrosion rate of Au coupled with Cu wires according to the mixed potential theory. According to the mixed potential theory, when Cu, Cu1Pd, and PC wire are coupled with Au in a galvanic couple, it is observed that the galvanic corrosion rate increases with an increase in HCl concentration. It is confirmed that the galvanic corrosion rate by the mixed potential theory is closely consistent with the electrochemical galvanic corrosion behavior.

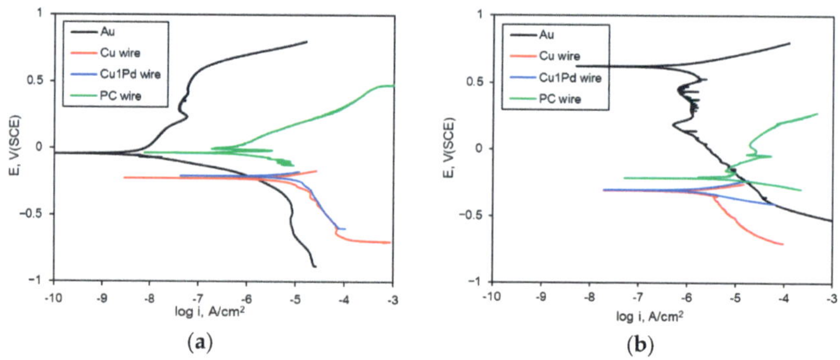

Figure 15. Analysis of galvanic corrosion using mixed potential theory between Au specimen and Cu wires; (**a**) 1% NaCl, (**b**) 1% NaCl + 0.1% HCl.

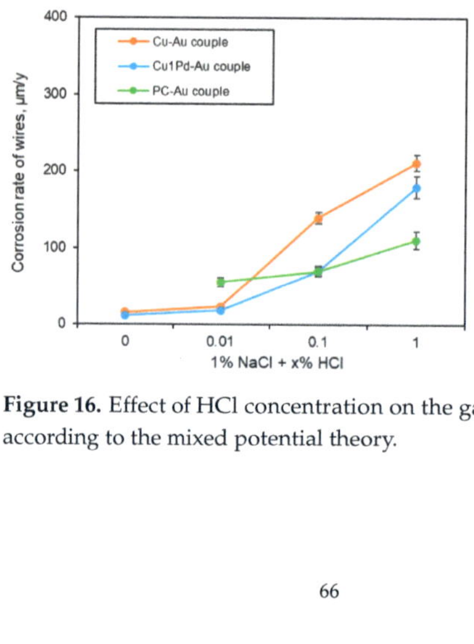

Figure 16. Effect of HCl concentration on the galvanic corrosion rate of Au coupled with Cu wires according to the mixed potential theory.

In semiconductor devices, galvanic corrosion between the bonding wires and bonding pads can significantly impact device reliability and performance over time. Initially, this may not have a significant effect on the device's performance, but exposure to electrolytes can trigger electrochemical reactions between the bonding wire and the pad. The resulting corrosion products increase the contact resistance, impeding the current flow and leading to power losses. As galvanic corrosion progresses, the physical bond may weaken, which can lead to micro-contact separation and circuit shorts or breaks. This degradation escalates device malfunction and failure rates, ultimately reducing device reliability.

If galvanic corrosion intensifies, it can sever electrical connections, leading to total device failure, with severe consequences such as data loss or system breakdowns. Galvanic corrosion can occur rapidly or develop slowly over an extended period. Therefore, factors such as environmental control, material selection, and surface treatments that prevent corrosion are critical considerations in the semiconductor manufacturing process.

The results and discussion presented above make it possible to propose the following galvanic corrosion mechanism between the bond pads and the wires (Figure 17): In the first ball bond area, corrosion occurs on the Al pad, which has a relatively active potential compared to the wires due to the galvanic effect, and Al oxide covers the pad and wire surroundings. However, the severe corrosive environment also causes the local corrosion of Cu, Cu1Pd, and PC wires. Meanwhile, in the second stitch bond area, galvanic corrosion occurs due to the relatively active potential of the wires compared to the Au pad, which preferentially corrodes the wire. The Cu and Cu1Pd wire dissolve and form an oxide covering the Au pad. Meanwhile, in the case of Pd-coated PC wire, Cu is only exposed to the pressed part, and the area of Cu acting as an anode becomes small due to the presence of the noble metal Pd on the wire surface. In this case, galvanic corrosion is further accelerated by the area effect of the small anode combined with the large cathode effect.

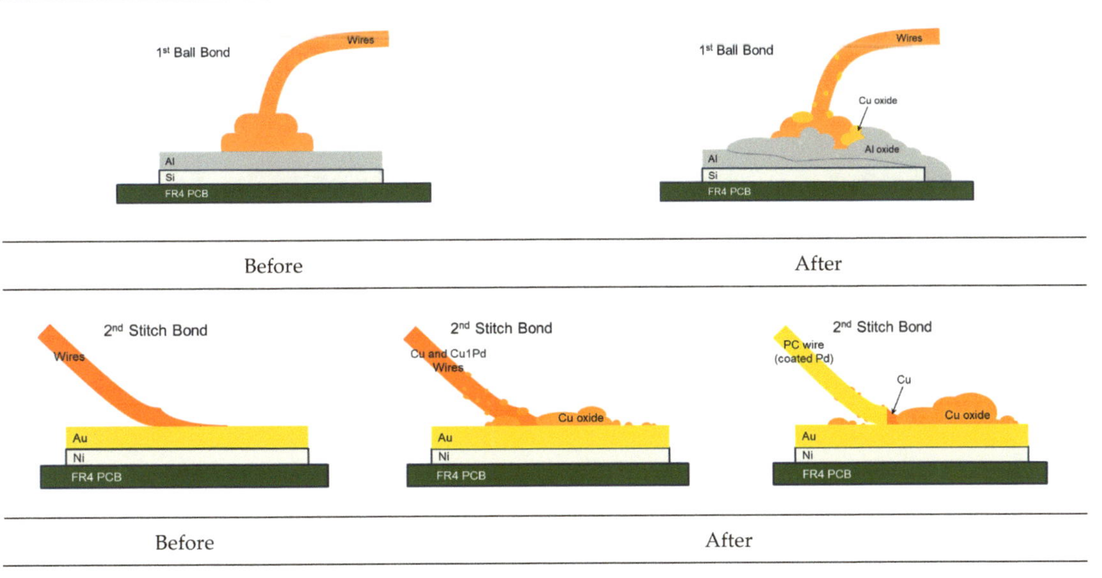

Figure 17. Proposed model for galvanic corrosion behavior between the bond pad and the wires on the 1st ball bond area and the 2nd stitch bond area.

4. Conclusions

In this study, to evaluate the influence of chloride, which is a contaminant formed during the PCB manufacturing process, the galvanic corrosion of bonding wire materials Cu wire, Cu wire alloyed with 1% Pd, and Pd-coated Cu wire was evaluated. The first ball bond was bonded to the Al pad and the second stitch bond was bonded to the Au pad in the preparation process, and the galvanic corrosion behavior of the semiconductor packaging module specimen was analyzed, resulting in the following conclusions;

1. In the first ball bonding area between the Al pad and the wires, the Al pad, which is a relatively active metal compared to the wires, undergoes preferential corrosion. For the same materials, galvanic corrosion between the Al pad and the three types of wires in terms of HCl concentration shows a high correlation with the anodic and cathodic Tafel constants;

2. In the second stitch bonding area between the Au pad and the wires, for the Cu and Cu1Pd wires, corrosion occurs in the wire because the Au pad has a relatively noble potential, while the formed Cu oxide is overlaid on the surface of the Au pad. On the other hand, in the Pd-coated PC wire, Cu is only exposed in the pressed area between the wire and the Au pad, and the galvanic corrosion is accelerated due to the small anode–large cathode effect because of the noble potential of the coated Pd and Au pad.

Author Contributions: Conceptualization, Y.-R.Y.; methodology, Y.-R.Y.; validation, Y.-R.Y.; formal analysis, Y.-S.K.; investigation, Y.-R.Y.; resources, Y.-S.K.; data curation, Y.-R.Y.; writing—original draft preparation, Y.-R.Y.; writing—review and editing, Y.-S.K.; visualization, Y.-R.Y.; supervision, Y.-S.K.; project administration, Y.-R.Y. and Y.-S.K.; funding acquisition, Y.-S.K. All authors have read and agreed to the published version of the manuscript.

Funding: This research was supported by a grant from the 2023–2024 research funds of Andong National University, and the Korea Institute for Advancement of Technology (KIAT) grant funded by the Korea Government (MOTIE) (HRD Program for Industrial Innovation).

Institutional Review Board Statement: Not applicable.

Informed Consent Statement: Not applicable.

Data Availability Statement: Data are contained within the article.

Acknowledgments: The authors would like to thank MK electron Co., Ltd., for supplying the materials.

Conflicts of Interest: The authors declare no conflicts of interest.

References

1. Cheng, P.Y.; Lai, P.Y.; Ye, Z.J.; Hsieh, C.L.; Ye, J.M. Effect of Pd distribution on Pd-plated Cu wire using different electronic flame off settings. *J. Mater. Sci. Mater. Electron.* **2017**, *28*, 4613–4618. [CrossRef]
2. Uno, T. Bond reliability under humid environment for coated copper wire and bare copper wire. *Microelectron. Reliab.* **2011**, *51*, 148–156. [CrossRef]
3. Qin, I.; Yauw, O.; Schulze, G.; Shah, A.; Chylak, B.; Wong, N. Advances in Wire Bonding Technology for 3D Die Stacking and Fan Out Wafer Level Package. In Proceedings of the 2017 IEEE 67th Electronic Components and Technology Conference (ECTC), Orlando, FL, USA, 30 May–2 June 2017. [CrossRef]
4. Qin, W.; Anderson, H.; Anderson, T.; Chang, G.D. Corrosion Mechanisms of Cu Wire Bonding on Al Pads. In Proceedings of the 2018 IEEE 68th Electronic Components and Technology Conference (ECTC), San Diego, CA, USA, 29 May–1 June 2018. [CrossRef]
5. Wu, Y.; Barton, S.C.; Lee, A. Galvanic corrosion behavior at the Cu-Al ball bond interface: Influence of Pd addition and chloride concentration. *Microelectron. Reliab.* **2019**, *92*, 79–86. [CrossRef]
6. Chauhan, P.; Zhong, Z.W.; Pecht, M. Copper Wire Bonding Concerns and Best Practices. *J. Electron. Mater.* **2013**, *42*, 2415–2434. [CrossRef]
7. Soestbergen, M.V.; Mavinkurve, A.; Zaal, J.J.M.; O'Halloran, G.M.; Rongen, R.T.H.; Farrugia, M.L. Crevice Corrosion of Ball Bond Intermetallics of Cu and Ag Wire. In Proceedings of the 2016 IEEE 66th Electronic Components and Technology Conference (ECTC), Las Vegas, NV, USA, 31 May–3 June 2016. [CrossRef]
8. Kumar, G.I.A.; Alptekin, J.; Caperton, J.; Salunke, A.; Chyan, O. Accelerated reliability testing of Cu-Al bimetallic contact by a micropattern corrosion testing platform for wire bond device application. *MethodsX* **2021**, *8*, 101320. [CrossRef]
9. Ross, N.; Asokan, M.; Kumar, G.I.A.; Caperton, J.; Alptekin, J.; Salunke, A.S.; Chyan, O.M. Mechanistic study of copper wire-bonding failures on packaging devices in acidic chloride environments. *Microelectron. Reliab.* **2020**, *113*, 113917. [CrossRef]
10. Fu, S.W.; Lee, C.C. A corrosion study of Ag-Al intermetallic compounds in chlorine-containing epoxy molding compounds. *J. Mater. Sci. Mater. Electron.* **2017**, *28*, 15739–15747. [CrossRef]
11. Ly, N.; Xu, D.E.; Song, W.H.; Mayer, M. More uniform Pd distribution in free-air balls of Pd-coated Cu bonding wire using movable flame-off electrode. *Microelectron. Reliab.* **2015**, *55*, 201–206. [CrossRef]

12. Kumar, G.I.A.; Lamber, A.; Caperton, J.; Asokan, M.; Yi, W.; Chyan, O. Comparative Study of Chloride and Fluoride Induced Aluminum Pad Corrosion in Wire-Bonded Device Packaging Assembly. *Corros. Mater. Degrad.* **2021**, *2*, 447–460. [CrossRef]
13. Chiu, Y.T.; Chiang, T.H.; Chen, Y.F.; Yang, P.F.; Huang, L.; Lin, K.L. The Corrosion Performance of Cu Alloy Wire Bond on Al Pad in Molding Compounds of Various Chlorine Contents under Biased-HAST. In Proceedings of the 2014 IEEE 64th Electronic Components and Technology Conference (ECTC), Lake Buena Vista, FL, USA, 27–30 May 2014. [CrossRef]
14. Soestbergen, M.V.; Ernst, L.J.; Zhang, G.Q.; Rongen, R.T.H. Transport of Corrosive Constituents in Epoxy Molding Compounds. In Proceedings of the 2007 International Conference on Thermal, Mechanical and Multi-Physics Simulation Experiments in Microelectronics and Micro-Systems. EuroSime 2007, London, UK, 15–18 April 2007. [CrossRef]
15. Tai, C.T.; Lim, H.Y.; Teo, C.H.; Audrey Swee, P.J. An investigation on Cu Wire Bond Corrosion and Mitigation Technique for Automotive Reliability. In Proceedings of the 2012 35th IEEE/CPMT International Electronics Manufacturing Technology Conference (IEMT), Ipoh, Malaysia, 6–8 November 2012. [CrossRef]
16. Kim, H.J.; Lee, J.Y.; Paik, K.W.; Koh, K.W.; Won, J.H.; Choe, S.H.; Lee, J.; Monn, J.T.; Park, Y.J. Effects of Cu/Al Intermetallic Compound (IMC) on Copper Wire and Aluminum Pad Bondability. *IEEE Trans. Compon. Packag. Technol.* **2003**, *26*, 367–374. [CrossRef]
17. Mathew, V.; Wikramanayake, E.; Chopin, S.F. Corrosion of Copper Wire bonded Packages by Chlorine Containing Foreign Particles. In Proceedings of the 2020 IEEE 70th Electronic Components and Technology Conference (ECTC), Orlando, FL, USA, 3–30 June 2020. [CrossRef]
18. Goh, C.S.; Chong, W.L.E.; Lee, T.K.; Breach, C. Corrosion Study and Intermetallics Formation in Gold and Copper Wire Bonding in Microelectronics Packaging. *Crystals* **2013**, *3*, 391–404. [CrossRef]
19. Liu, D.; Chen, H.; Wu, J.; Then, E. Corrosion Behavior of Cu-Al Intermetallic Compounds in Copper Wire Bonding in Chloride Containing Accelerated Humidity Testing. In Proceedings of the 2016 IEEE 66th Electronic Components and Technology Conference (ECTC), Las Vegas, NV, USA, 31 May–3 June 2016; pp. 629–636.
20. Yang, Y.; Zhu, X.; Nistala, R.R.; Zhao, S.; Xu, J. A method to simulate fluorine outgas and Al bond pad corrosion in wafer fab. In Proceedings of the 2016 IEEE 23rd International Symposium on the Physical and Failure Analysis of Integrated Circuits (IPFA), Singapore, 18–21 July 2016. [CrossRef]
21. Younan, H.; Xiang, X.Z.; Xiaomin, L. Characterization studies of fluorine-induced corrosion crystal defects on microchip Al bondpads using X-ray photoelectron spectroscopy. In Proceedings of the 2014 IEEE 21th International Symposium on the Physical and Failure Analysis of Integrated Circuits (IPFA), Singapore, 30 June–4 July 2014. [CrossRef]
22. Fu, C.H.; Hung, L.Y.; Jiang, D.S.; Chang, C.C.; Wang, Y.P.; Hsiao, C.S. Evaluation of new substrate surface finish: Electroless nickel/electroless palladium/immersion gold (ENEPIG). In Proceedings of the 2008 58th Electronic Components and Technology Conference, Lake Buena Vista, FL, USA, 27–30 May 2008. [CrossRef]
23. Zeng, Y.; Bai, K.; Jin, H. Thermodynamic study on the corrosion mechanism of copper wire bonding. *Microelectron. Reliab.* **2013**, *53*, 985–1001. [CrossRef]
24. Farrah, H.; Slavek, J.; Pickering, W. Fluoride interactions with hydrous aluminum oxides and alumina. *Soil Res.* **1987**, *25*, 55–69. [CrossRef]
25. Proost, J.; Baklanov, M.; Verbeeck, R.; Maex, K. Morphology of corrosion pits in aluminum thin film metallizations. *J. Solid State Electrochem.* **1998**, *2*, 150–155. [CrossRef]
26. Loto, R.T. Investigation of the Localized Corrosion Resistance of 4044 Aluminum Alloy in Acid Chloride and Neutral Chloride Solutions. *J. Fail. Anal. Prev.* **2018**, *18*, 905–911. [CrossRef]
27. Yoo, Y.-R.; Kim, G.; Jeon, S.-M.; Park, H.-J.; Seo, W.-W.; Moon, J.-T.; Kim, Y.-S. Influence of HCl Concentration on Corrosion Behavior between Au or Cu Bonding Wires and the Bond Pad for Semiconductor Packaging. *Materials* **2023**, *16*, 7275. [CrossRef] [PubMed]
28. Zhong, Z.W. Wire bonding using copper wire. *Microelectron. Int.* **2009**, *26*, 10–16. [CrossRef]
29. Lee, T.K.; Breach, C.D.; Chong, W.L.; Goh, C.S. Oxidation and corrosion of Au/Al and Cu/Al in wire bonding assembly, 2012. In Proceedings of the 13th International Conference on Electronic Packaging Technology & High Density Packaging, Guilin, China, 13–16 August 2012. [CrossRef]
30. Gan, C.L.; Ng, E.K.; Chan, B.L.; Hashim, U.; Classe, F.C. Technical Barriers and Development of Cu Wirebonding in Nanoelectronics Device Packaging. *J. Nanomater.* **2012**, *2012*, 173025. [CrossRef]
31. Hong, W.S.; Kim, M.-R.; Kim, S.Y.; Jeon, S.M.; Moon, J.T.; Kim, Y.-S. Corrosion characteristics of Gold-Coated Silver Wire for Semiconductor Packaging. *Corros. Sci. Technol.* **2021**, *20*, 289–294. [CrossRef]
32. Lim, A.B.Y.; Chang, A.C.K.; Yauw, O.; Chylak, B.; Gan, C.L.; Chen, Z. Ultra-fine pitch palladium-coated copper wire bonding: Effect of bonding parameters. *Microelectron. Reliab.* **2014**, *54*, 2555–2563. [CrossRef]
33. Wu, Y.; Subramanian, K.N.; Barton, S.C.; Lee, A. Electrochemical studies of Pd-doped Cu and Pd-doped Cu-Al intermetallics for understanding corrosion behavior in wire-bonding packages. *Microelectron. Reliab.* **2017**, *78*, 355–361. [CrossRef]
34. Lau, K.T.; Cha, C.L. A Review of Palladium Coated Copper Wire Bonding For Automotive Device. *Int. J. Mech. Prod. Eng. Res. Dev.* **2020**, *10*, 4479–4492.
35. Fontana, M.G. *Corrosion Engineering*, 3rd ed.; 1 International Business Park #01-15A; MCGraw-Hill Book company: Singapore, 1985; pp. 41–51, 463–468.

36. Kaimori, S.; Nonaka, T.; Mizoguchi, A. The development of Cu bonding wire with oxidation-resistant metal coating. *IEEE Trans. Adv. Packag.* **2006**, *29*, 227–231. [CrossRef]
37. Singh, I.; Qin, I.; Xu, H.; Huynh, C.; Low, S.; Clauberg, H.; Chylak, B.; Acoff, V.L. Pd-coated Cu wire bonding technology: Chip design, process optimization, production qualification and reliability test for high reliability semiconductor devices. In Proceedings of the 2012 IEEE 62nd Electronic Components and Technology Conference, San Diego, CA, USA, 29 May–1 June 2012. [CrossRef]
38. Leong, G.C.; Uda, H. Comparative reliability studies and analysis of Au, Pd-coated Cu and Pd-doped Cu wire in microelectronics packaging. *PLoS ONE* **2013**, *8*, e78705. [CrossRef] [PubMed]

coatings

MDPI

Article

Optical Limiting from CdSe-Based Multiphase Polymer Nanocomposite Films

Leah M. Eversole, Richard Adjorlolo, Jack Francis Renaud and Mithun Bhowmick *

Department of Mathematical and Physical Sciences, Miami University Regionals, Middletown, OH 45042, USA; eversolm@miamioh.edu (L.M.E.); adjorlr@miamioh.edu (R.A.); renaudjf@miamioh.edu (J.F.R.)
* Correspondence: bhowmim@miamioh.edu

Abstract: Closely packed nanoparticles in polymer films are interesting materials where collective as interactive optical properties could be tuned based on nanoparticle proximity, surface morphology, types of encapsulation and matrix parameters. Two types of polymers (polymethylmethacrylate (PMMA) and polyvinyl alcohol (PVA))-based nanocomposite films featuring dual-colored emission peaks (~578 nm and ~650 nm) were fabricated from CdSe quantum dots to study their viability in optoelectronic applications. Using a 405 nm excitation laser, the evolution of photoluminescence (PL) intensities and peak wavelengths were examined as a function of increasing excitation intensity. While PL intensities showed systematic saturation and quenching, the emission wavelengths were found to be linearly red shifting with increasing excitation intensities in the PMMA films. The 650 nm emitting QDs seem to tune the PL saturation behavior in these films, as opposed to the PVA-based materials, where no such impact was seen. The material system could be a low-cost, low-maintenance alternative for future mesoscale sensing and light-emitting device applications.

Keywords: quantum dots; polymethyl methacrylate; polyvinyl alcohol; photoluminescence; polymer nanocomposites; CdSe

Citation: Eversole, L.M.; Adjorlolo, R.; Renaud, J.F.; Bhowmick, M. Optical Limiting from CdSe-Based Multiphase Polymer Nanocomposite Films. *Coatings* **2024**, *14*, 634. https://doi.org/10.3390/coatings14050634

Academic Editor: Xu Long

Received: 28 March 2024
Revised: 8 May 2024
Accepted: 16 May 2024
Published: 17 May 2024

1. Introduction

Nanoscience in physics and materials science primarily delves into understanding optoelectronics at nm-sized dimensions, while in chemistry, these applications are associated with several different systems, including but not limited to colloids, micelles, polymer composites, and other similar structures [1–9]. This overlapping of disciplines enables the engineering of novel materials with size-dependent optical properties, opening new avenues such as cybernetics, non-invasive clinical procedures, lab on a chip, and many environmental applications [9–19]. Semiconductor nanomaterials have attracted a lot of attention due to their widespread applicability in optoelectronics. Substantial efforts have been invested in the synthesis and characterization of semiconductor-based nanomaterials from groups II–VI, III–V, III–VI, and oxide-semiconductor materials until now [1]. Among these, a special class of materials are wide band gap II-VI semiconductors, since their bulk properties are within the visible spectrum and their applications have a wide range [1–3].

Quantum dots (QDs) are sometimes utilized in diverse and flexible applications across various environments, landscapes, and geometries [9,10]. Sometimes, using QDs directly in the application site can result in aggregation, potentially resulting in photodegradation [5,11]. One viable solution is the use of composite nanomaterials. Composite nanomaterials, which consist of more than one component, are employed in situations requiring flexibility and several of the properties coming from the components that are present. In many instances, QDs are dispersed within polymer matrices, forming a polymer nanocomposite (PN) [5,13]. In PN systems, the nanoparticles exhibit strong repulsion, thereby reducing the likelihood of aggregation [13]. PNs are particularly advantageous for designing novel materials due to their lightweight nature and ease of processing. Recent theoretical calculations provide accurate predictions of the optical properties of PNs, offering a strong platform where predictive models enhance the optical properties of both the polymer and the QDs. The models include details of the chemical properties of the polymer matrices and the confinement-related properties of the nanoparticles [8–11].

Photoluminescence (PL) stands out as one of the most powerful fundamental properties of PN systems. Upon absorbing a suitable photon, an electron transitions into a higher energy state, before re-emitting another photon to relax back into the ground state [4]. The emitted photons typically possess lower energies or longer wavelengths because of Stoke's shift [4]. Although the basic mechanism remains the same, the specifics of the electronic transitions, metastable states, and relaxation processes are influenced by various factors, such as scattering, defect, or impurity states. The mechanism of detailed balance is also responsible in certain cases [2]. Therefore, characterizing photoluminescence (PL) from PNs becomes crucial, especially when they are in different chemical environments and/or subjected to varying interfacial interactions.

A wide range of applications, including but not limited to light-emitting devices, solar cells, sensors, nano-catalysts, greener memories, and engineering of tissues, has been achieved through II–VI PN materials [1,5–7,10,14]. CdSe PNs hold a special place among them and have been attractive for their suitability in synthesis recipes as a stable nanomaterial [14]. CdSe QDs typically consist of a CdSe core and a ligand shell. Ligands play crucial roles in maintaining the stability and solubility of the nanoparticles. Also, during the synthesis process, ligands play a stabilizing role by preventing the aggregation and precipitation of the nanocrystals, thus ensuring controlled growth. For all practical purposes, it is important to probe the effects of closely packed CdSe nanoparticles on their optical properties.

This work focuses on the synthesis of polymer materials doped with CdSe QDs exhibiting two distinct PL emission wavelengths. PL intensities from the two characteristic wavelengths, originating from QDs of different sizes within the system, were adjustable by varying the ratio of doped silica to polymethyl methacrylate (PMMA) microparticles. The objective of this study was to probe the effects of varying orange to red-emission peak intensity ratios on the PL emission intensities as the excitation intensity increases. The measurements captured the PL intensity growth when the laser excitation intensity was modified, by measuring the PL emitted from the two-colored QD systems. Examination of the data unveiled distinct trends in the growth of PL intensity concerning laser excitation when compared to similar CdSe PNs made with polyvinyl alcohol (PVA) matrix, as discussed herein [12]. The importance of this work could be noted: (a) in the uniqueness of the nanocomposite systems created and studied, (b) in the way their steady-state PL responds to the excitation intensity change, and (c) due to the link between the PL growth response to the relative amount of "red" dots present. Essentially, this report emphasizes the synthesis of a nanocomposite system that is easy to integrate and has unique optical limiting properties.

2. Materials and Methods

QD films with two nanoparticle sizes were generated by initially suspending the desired quantity of silica microparticles doped with "orange" QDs, which emit at around 578 nm, in 2.3 g of a 5% polymethyl methacrylate (PMMA) solution. To create a 50% silica film, 0.1 g of microparticles was incorporated, and for a 25% silica film, 0.04 g of microparticles was added, and so forth. A detailed description documented in an earlier report focused on PVA-based material, and hence only a brief overview will be included here [10,12]. First, the doping of silica microparticles with QDs was performed. This was achieved by creating a suspension of 0.1 g of commercially available silica microparticles (Sigma Aldrich, St. Louis, MO, USA) in 3 mL of stock solution of ammonia after mixing it with 4 mL of ethanol and 2 mL of distilled water. The mixture was stored in a 20 mL vial. Then, 10 mg of "orange" emitting quantum dots (QDs) were added to this solution before treating it with a surface ligand (3-Mercaptopropyl) trimethoxysilane (purchased from

Mesolight, Inc., Suzhou, China), with 0.4 g of tetraethyl orthosilicate (TEOS). At this point, the vial was sealed, before stirring the mixture for an hour at temperature. The reaction mixture was allowed to settle next. Once settled, the coloration of the supernatant was checked. In case there is a coloration, an extra amount (0.1 g) of TEOS is added, before repeating the stirring step for one more hour. The supernatant was made transparent by repeating the above two steps a few times. Next, 0.1 g of aminopropyltriethoxysilane (APTES) and 0.1 g of distilled water were added, and the reaction vessel was stirred for 2 h. The reaction product was centrifuged to gather the particles, before washing with isopropanol, and finally drying them under vacuum. After making a solution comprising microparticles and the polymer, careful mixing was carried out to achieve even dispersion. At this point, "red" QDs with PL at approximately 650 nm were added to the solution in instalments of 10 μL and constantly mixed. After adding each instalment of "red" QDs, a droplet of the resulting product was checked under a microscope while being vacuum-dried. Throughout this text, the terms "orange" and "red" PL will occasionally be denoted as "peak 1" and "peak 2", respectively. It is noteworthy that the difference in the fabrication process from one sample to another is only the relative amount (in volume) of the "red" (650 nm) vs. "orange" (578 nm) emitting QDs. Essentially, when "red" emitting QDs were added by volume in instalments of 10 μL, after each instalment the relative PL intensities were checked. This process was stopped once the desired PL peak ratio was achieved.

Once the desired PL peak ratios were achieved, the solutions were drop-cast on 50.8 mm × 50.8 mm × 6.35 mm glass substrates (Chemglass Life Sciences LLC., Vineland, NJ, USA). The glass substrates were chosen based on their high transmittance in the wavelength range 300–2000 nm. During the drop cast process, ~5 μL of the solution was pipetted on the glass substrate and the solution was spread using a spatula to have a uniform film. Once the films were dried, the drop cast samples were inspected using regular microscope (AmScope LLC., Irvine, CA, USA). Using surface profilometry (Dektak 3ST from Bruker Corporation, Billerica, MA, USA) the thicknesses of the films were found to be ~90 μm with 5% or less variance. The PL peak position and intensity did not show any dependence on the thickness of the layer. Figure 1a illustrates a schematic cross-section of the sample, showing that the "orange" QDs are connected to silica, while the "red" QDs are evenly spread in the polymer matrix. The ratio was varied to probe the effect of the weak and strong presence on the interaction between the two QDs. The experimental setup is presented in Figure 1b. A closeup of the deposited PMMA1 film is shown in Figure 1c, showing reasonable smoothness on the surface. A scanning electron microscope (SEM) from Zeiss (White Plains, NY, USA) was used to comment on the texture and surface of the film, as could be seen in Figure 1d. The SEM confirmed the presence and uniformity of the deposited film. While no characterization of surface contamination could be made, it is safe to assume that the level of contamination has not impacted the overall PL signal, which is the focus of this study.

The comparative intensities from both PL peaks were investigated by exciting samples to a 405 nm diode laser with tunable power control. Table 1 outlines the 6 samples examined in this study, where PMMA1 refers to an "orange":"red" intensity ratio of 10:1, PMMA2 refers to 8:1, and so on, as detailed in Table 1. The two PVA-based samples studied were PVA1 and PVA2, with the "orange": "red" peak intensity ratios 4:1 and 2:1, respectively. The ratios were chosen (1) to probe the effects of the "orange":"red" PL peak intensity ratio on the growth of PMMA-based samples, and (2) to compare PMMA and PVA-based samples with similar peak intensity ratios. PMMA3 and PVA1 share the same intensity ratio, and thus they are comparable for a preliminary investigation of the role of polymer in the PL properties. Similarly, PMMA4 and PVA2 are similar and hence have been compared in the discussion.

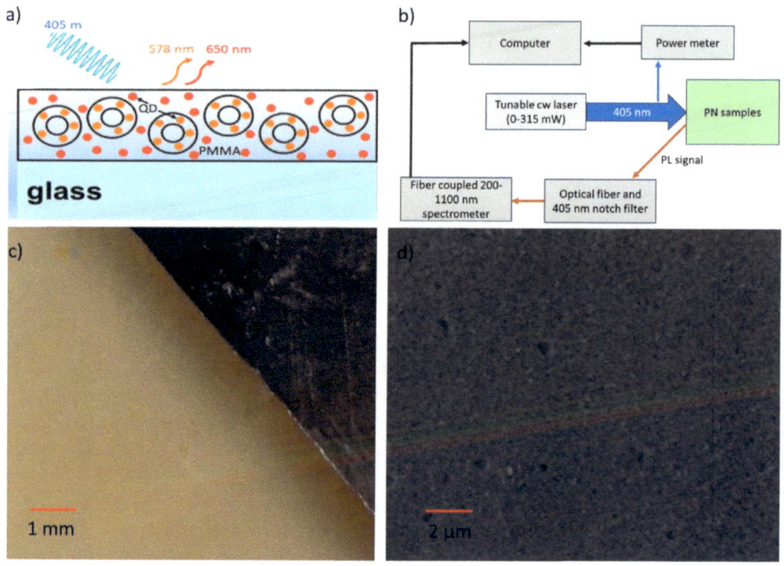

Figure 1. A schematic diagram of the cross-section of two-phase two-QD samples showing orange QDs bound in the shells of silica microparticles and "red" QDs dispersed in the surrounding polymer (PMMA) matrix (**a**). The "red" dots are representative of the 650 nm emitting QDs, whereas the "orange" dots are 578 nm emitting QDs encapsulated inside the silica microparticles (black concentric circles). The films were drop-cast on glass substrates. The experimental setup showing PL measurement scheme (**b**). A photograph of PMMA1 film deposited on glass, with an aluminum foil placed next to it for contrast (**c**), and an SEM image (**d**) of PMMA1 film.

Table 1. List of CdSe PNs synthesized and investigated.

Samples	Peak 1 and 2 Intensity Ratios at 405 nm, 0.014 W/cm^2 Intensity
PMMA1	10:1
PMMA2	8:1
PMMA3	4:1
PMMA4	2:1
PVA1	4:1
PVA2	2:1

The PL was measured using a custom-built setup employing backscattering geometry, featuring a fiber-coupled spectrometer (Silver Nova from StellerNet Inc., Tampa, FL, USA). A schematic diagram is presented in Figure 1b. A 405 nm cw laser was used as the excitation source. The excitation laser intensities could be tuned through a calibrated average power vs. diode current curve, spanning from 0 to 315 mW with 3% or less fluctuations, monitored by a power meter from Laserglow Technologies (North York, ON, Canada). The laser powers were subsequently converted to intensities using the following formula:

$$\text{intensity (W/cm}^2) = \text{output power (W)/area of laser beam (cm}^2) \quad (1)$$

where, "output power" is defined as the measured power at the sample plane, while "Area" denotes the region covered by the laser beam at that plane. Unless specified otherwise, the exposure time was set at 5 s, and the beam area was estimated to be 0.1257 cm^2 for all reported measurements. PL was collected using a fiber-coupled assembly fitted with a 405 nm notch filter to separate the excitation signal from the PL. All experiments were carried out at room temperature. Consequently, photoluminescence (PL) measurements were performed, varying the 405 nm laser intensity within the range of 0–2.5 W/cm^2.

All four samples were found to have stable and consistent PL at the respective peak wavelengths. A series of 10 measurements were made stretched over an interval of 1 minute confirming the stability and consistency of the samples, as presented in Figure 2a,b. It is clear from Figure 2a that the two peaks are reproducible and there is no significant variability in intensities for the 1-minute interval of time, which is much greater than the 5 s interval for which the samples were excited in the PL measurements in this work. In Figure 2b, the intensity and wavelength consistency are plotted vs. the number of "shots". Each of the 10 measurements is referred to as "shots" in that plot. Albeit insignificant, the wavelength from peak 2 showed more variability (5% or less) relative to peak 1 (4% or less). All the PMMA-based samples reported here demonstrated similar consistencies. The PVA-based samples were also identical, as reported previously [12]. The measurements reported here have been repeated on multiple occasions on different days to ensure that the measurements are consistent, the changes in PL properties are reversible, and to rule out any degradation impacting the optical properties.

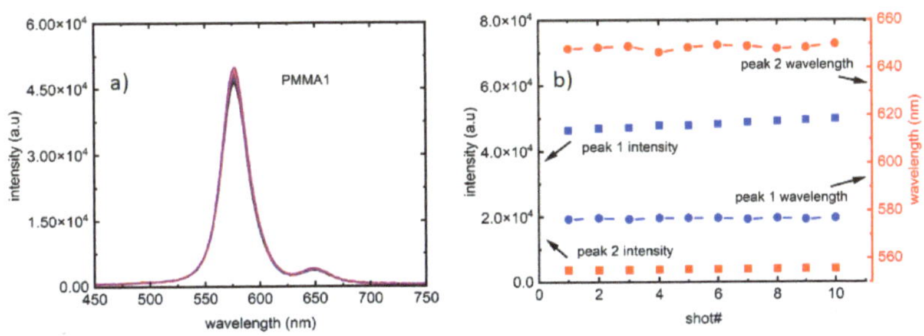

Figure 2. PL repeatability test for intensity and wavelength. (**a**) A total of 10 measurements (termed as "shots") collected and overlaid for PMMA1, where different lines present the degree of fluctuations between shots within a period of 1 minute, which is 4% or less. (**b**) PL wavelength and intensity consistency for the two peaks for PMMA1 for the 10 shots.

3. Results

Once the initial characterization was complete, a systematic study of the PL growth was undertaken. Figure 3 presents the PL measurements from the four samples and the analysis. Figure 3a presents the four PMMA-based samples at a laser intensity of 0.2275 W/cm^2 excitation. The intensity ratios listed in Table 1 are for this excitation level. The ratios are dependent on not only individual peaks, but also their relative strengths, which is noteworthy, and not obvious from Figure 3a. The intensity growth in the PMMA-based samples seems to be sensitive to the presence of "red" quantum dots, emitting at 650 nm (also referred to as peak 2 in this work). Figure 3b–d present the growth of the two emission peaks as a function of laser excitation intensities. A comparison between the two peaks for PMMA1 is presented in Figure 3b as an example. While both peaks show similar trends in initial linear increase and eventual saturation, there are differences in thresholds of saturation, and the rapidity of intensity growth in them. This trend is seen in all PMMA samples, where the peaks grew almost linearly, after which a saturation of PL sets in. However, PMMA1 showed a larger saturation threshold for both peak 1 (~1.5 W/cm^2) and peak 2 (~2 W/cm^2), compared to other PMMA-based samples, where the saturation starts at ~1 W/cm^2. The intensities at which PMMA1 peaks saturated are also significantly higher than the other samples. For example, saturation level PL intensity in PMMA1 for peak 1 is ~4 times that of PMMA2 and PMMA3. Compared to PMMA4, PMMA1 is more than an order of magnitude higher in the saturating PL intensity for both peaks.

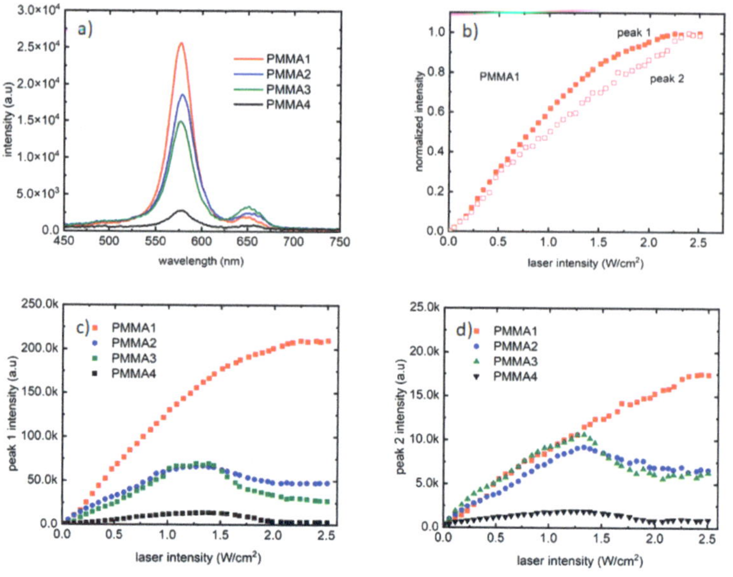

Figure 3. Fluorescence from the four samples as the excitation laser intensity varied from 0–2.5 W/cm^2. (**a**) The four PMMA-based samples studied, (**b**) evolution of PL peak 1 and 2 in PMMA1 with laser intensity, (**c**) PL peak 1 PL intensity growth in PMMA-based samples, (**d**) PL peak 2 PL intensity growth in PMMA-based samples.

Another important distinction is the presence of PL quenching in all PMMA-based samples except PMMA1. Figure 3c,d show clearly the reduction in peak intensities in PMMA2, PMMA3, and PMMA4, shortly after the saturation takes place. There is no quenching present in PMMA1. These features are unique, since in a previous report on PVA-based PN samples, there was no saturation reported for the 578 nm peak (or peak 1), while the 650 nm peak showed slight saturation at ~2.5 W/cm^2. This difference between PMMA-based and PVA-based PN can be seen readily in Figure 4, where the evolution of the two peaks was compared in two similar sets of samples. In Figure 4a, the 578 nm peak intensities are compared for two PMMA-based samples with two PVA-based samples. The peak intensity ratio in PMMA3 and PVA1 is 4:1, while the same for PMMA4 and PVA2 is 2:1. A similar comparison is presented for the 650 nm peak in Figure 4b. The difference in PMMA-based and PVA-based samples could be readily seen. Both PVA-based samples continued to grow linearly as a function of excitation intensity, and there is no trace of saturation for peak 1. The 650 nm peak (or peak 2) did saturate in PVA-based samples. However, the threshold of saturation is significantly lower in PMMA samples (~1 W/cm^2) compared to the PVA samples (~2.5 W/cm^2). Note that the general trend for PVA-based samples presented here is consistent with previous studies, irrespective of the peak intensity ratios [12].

Another aspect of the multiphase PL emissions is the peak shift as a function of laser excitation intensity. In a recent report, the PVA-based PNs have been found to emit at longer wavelengths at higher excitation intensities [12]. Figure 5 illustrates this through a comparison between PMMA and PVA-based samples, capturing the redshift for both peaks. Each symbol in Figure 5 presents an average of the peak positions found from all four samples, with the error bars showing standard deviation from them. There is hardly much difference in the redshifts of the two types of samples, except for slightly steeper shifts in PMMA-based samples, and an overall larger uncertainty, expressed as one standard deviation in the error bars.

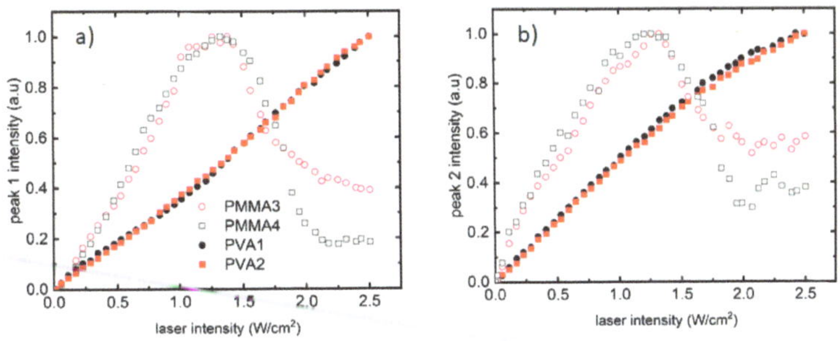

Figure 4. Comparison of the two PL peak intensity growths in PMMA-based (open symbols) and PVA-based (solid symbols) samples. (**a**) Comparison for the 578 nm peak and (**b**) the same for the 650 nm peak.

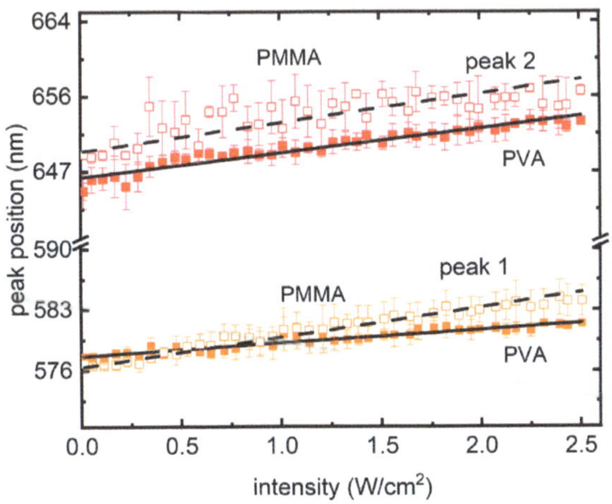

Figure 5. The shift in emission wavelength for the two peaks with increasing excitation intensity for PMMA (open symbols) and PVA (solid symbols)—based samples with similar peak intensity ratios where the error bars showing the standard deviation. The lines present linear fits to the peak positions for PMMA (dashed line) and PVA (solid line)—based materials.

A redshift in PL corresponds to a shrinkage of the band gap, which is different from single-phase QD films caused by dynamic band filling to increase the gap, resulting in a blueshift in PL peaks in those materials [11]. The linear redshift (goodness of fit > 96%) seen here is also different from previously reported nonlinear blueshift in PL peak wavelength, and has been modelled successfully using photo-dynamic Burstein–Moss theory [11].

4. Discussion

There are several major observations found in this study, such as: (1) a tunability in saturation intensities of both PL peaks, controlled by the relative peak intensity ratios, (2) dependence of PL-saturation on the amount of 650 nm QDs, (3) saturation and quenching in PMMA-based PNs, which was absent in PVA-based PNs, and (4) a redshift in both PL peaks as excitation intensity increased, a similar trait found in both types of PNs.

The tunability of saturation intensities, as shown in Figure 3c,d, is direct evidence that the 650 nm emitting QDs could control saturating PL intensities as well as corresponding excitation strengths needed. However, for "orange":"red" ratios $\leq$ 8:1, the samples also showed PL quenching when the excitation intensity grew higher. Saturation in QD systems has been reported previously and was attributed to the formation of a maximum number of electron–hole (e–h) pairs possible to be created [2,7,12]. However, quenching was not seen in the previous studies. One of the most common contributors to any quenching mechanism in colloidal nanoparticle systems is Auger recombination, where a non-radiative energy dissipation contributes to the consumption of energy of the e-h pairs in radiative systems [19]. Hence, it is very likely that the different quenching in PL shown by the PMMA samples is a result of Auger recombination, which should be an efficient process in multi-carrier systems such as the ones being studied here. Auger recombination is a complex process, with multiple pathways possible for the systems to take depending on the core–shell geometries, the inter-particle distances, the dimensions of the dots and the surface states available in the systems [19].

It is very likely that a charge transfer mechanism is also contributing to the optical limiting behavior in the samples. This would explain the role of relative peak intensities on the overall saturation behavior. Clearly, when more 650 nm dots are available to participate in interdot optical exchange processes, more impact would be seen on the PL properties. Polymer-based multiple-sized CdSe QD systems were previously reported to have charge transfer processes present, resulting in quenching [20]. However, there was no prior evidence for quenching controlled by the choice of polymers. As shown in Figure 4, both QDs behaved differently when they were embedded in PMMA vs. PVA. The difference is most likely an attribute of nucleation in PVA resulting from water porosity which translates into changes in surface morphology and stiffness. It has been shown that PVA-based carbon dot PN system has differences in PL behavior because of the inherent properties of the polymer, that could be used to overcome the PL saturation in them [21]. This could explain the continuous growth in PVA-based PNs, vs. saturation and quenching in PMMA materials, which are not as porous, as seen in Figure 4. The morphology and surface roughness in PN systems significantly contribute to the optical properties. It is entirely possible that the surface morphologies of the two different polymers are partially responsible for the difference in optical losses, as was reported previously [22].

Both peaks almost linearly redshifted as the laser excitation intensity increased. However, PMMA-based samples showed larger sensitivity by shifting ~7 nm, compared to a marginal ~3 nm shift in PVA-based PNs at the highest laser excitation intensity of 2.5 W/cm^2. This is clearly a local stress-dependent property that is common in high-pressure PL properties [5]. Usually, compression allows the band gap to shrink in materials, making them emit at longer wavelengths, as seen to be happening in Figure 5. The red-shift could be quantified as a local stress metric, and hence the PNs could be utilized in pressure-sensing devices, as ultra-sensitive shock wave detectors.

The saturation and quenching behavior demonstrated by PN systems in this work could be classified as optical limiting, where the transmission of the material changes at higher laser intensity. This property could be employed in an all-optical switch. Unlike the PVA-based materials, where the PL peak intensities grow indefinitely and linearly, the saturation and subsequent quenching in PMMA-based materials make them extremely versatile with more than one, optically controllable PL peak intensity level.

5. Conclusions

The main purpose of this study is to highlight applications related to optical limiting properties in a PN system comprising CdSe and PMMA and to highlight the differences they have when compared to PVA-based similar systems. The PN system is easy to integrate, and partly linear and partly nonlinear behavior in its PL response when optical excitation is systematically increased.

All PMMA-based samples showed saturation. However, the saturation thresholds are different. PMMA1 saturated at ~1.5 W/cm^2 of laser intensity, while PMMA2 and PMMA3 saturated at ~1 W/cm^2. The two peaks, i.e., "orange" and "red" emission peaks evolved similarly with increasing laser excitation intensities. Interestingly, the PMMA-based samples showed photoinduced quenching as the excitation laser intensity reached 1.5 W/cm^2 and above, which was not present in the PVA-based samples of similar "orange":"red" PL peak intensity ratios. At ~2 W/cm^2, the PMMA-based samples showed a second saturation of PL peak intensity, this time at a lower value. All photoinduced effects reported here are reversible, and the processes are nondestructive. PMMA-based samples showed better sensitivity to local stress compared to PVA-based samples by more than twice as much redshift.

A complete understanding of the mechanism in the evolution of PL in the four samples presented here seems complex and could be dependent on several factors. However, certain general observations could be made. The PMMA-based nanocomposites synthesized and characterized through PL here were found to be stable, consistent photoemitters with their peak 1 vs. peak 2 ratios staying the same for a large range of excitation intensities from a 405 nm laser. In contrast to previously reported PVA-based materials, the PMMA-based samples showed distinguishing saturation and quenching. The PNs showed consistent stress-dependent PL peaks, which are valuable as multiphase pressure-sensitive devices. The QD samples appear resilient to photo-induced permanent degradation which has been a critical issue in inorganic QD-based solar cells and light-emitting devices in the recent past [15–18]. Auger recombination and charge transfer could be the two major contributing factors in the observed PL saturation and quenching trends. The measurements reported here are exciting for future light-emitting device applications. The samples studied here use a simple recipe, making them a high-throughput synthesis process. Their PL intensity is substantial even in a thick film formation. They are easy to combine into different morphologies and shapes due to the inherent flexibility coming from the polymers.

Author Contributions: Conceptualization, M.B.; methodology, M.B.; formal analysis, M.B., L.M.E., J.F.R. and R.A.; investigation, L.M.E., M.B. and R.A.; resources, M.B.; data curation, M.B., J.F.R. and R.A.; writing—original draft preparation, M.B.; writing—review and editing, M.B.; supervision, M.B.; project administration, M.B. All authors have read and agreed to the published version of the manuscript.

Funding: This research received no external funding.

Institutional Review Board Statement: Not applicable.

Informed Consent Statement: Not applicable.

Data Availability Statement: The data presented in this study are available within the article.

Acknowledgments: M.B. thanks Bruno Ullrich for his valuable comments on the analysis.

Conflicts of Interest: The authors declare no conflict of interest.

References

1. Klimov, V.I. (Ed.) *Nanocrystal Quantum Dots*, 2nd ed.; CRC Press: New York, NY, USA, 2010.
2. Bhowmick, M.; Ullrich, B.; Murchland, M.; Zhou, X.; Ramkumar, C. Substrate and Excitation Intensity Dependence of Saturable Absorption in Perovskite Quantum Dot Films. *Nanomaterials* **2023**, *13*, 871. [CrossRef] [PubMed]
3. Moreels, I.; Kruschke, D.; Glas, P.; Tomm, J.W. The dielectric function of PbS quantum dots in a glass matrix. *Opt. Mater. Express* **2012**, *2*, 496. [CrossRef]
4. Ullrich, B.; Antillón, A.; Bhowmick, M.; Wang, J.S.; Xi, H. Atomic transition region at the cross-over between quantum dots to molecules. *Phys. Scr.* **2014**, *89*, 025801. [CrossRef]
5. Christensen, J.M.; Banishev, A.A.; Dlott, D.D. Bright emissive core-shell spherical microparticles for shock compression spectroscopy. *J. Appl. Phys.* **2014**, *116*, 033513. [CrossRef]
6. Kang, Z.; Banishev, A.A.; Lee, G.; Scripka, D.A.; Breidenich, J.; Xiao, P. Exploration of CdTe quantum dots as mesoscale pressure sensors via time-resolved shock-compression photoluminescent emission spectroscopy. *J. Appl. Phys.* **2016**, *120*, 043107. [CrossRef]
7. Bhowmick, M.; Singh, A.K.; Barik, P.; Xi, H.; Ullrich, B. All-optical switch based on PbS quantum dots. *Appl. Phys. Lett.* **2021**, *119*, 192103. [CrossRef]
8. Ullrich, B.; Xiao, X.Y.; Brown, G.J. Photoluminescence of PbS quantum dots on semi-insulating GaAs. *J. Appl. Phys.* **2010**, *108*, 013525. [CrossRef]
9. Zimnitsky, D.; Jiang, C.; Xu, J.; Lin, Z.; Tsukruk, V.V. Substrate- and Time-Dependent Photoluminescence of Quantum Dots Inside the Ultrathin Polymer LbL Film. *Langmuir* **2007**, *23*, 4509–4515. [CrossRef] [PubMed]
10. Christensen, J.M. Fast Photoemissive Probes for Shock Compression Microscopy. Ph.D. Dissertation, University of Illinois, Urbana, IL, USA, 2019. Available online: http://hdl.handle.net/2142/104819 (accessed on 4 January 2024).

11. Ullrich, B.; Barik, P.; Singh, A.K.; García-Ramírez, E.V.; Reyes-Esqueda, J.A. Photo-dynamic Burstein-Moss doping of PbS quantum dots in solution by single and two-photon optical pumping. *Opt. Mater. Express* **2015**, *5*, 2431–2436. [CrossRef]
12. Bhowmick, M.; Christensen, J.; Adjorlolo, R.; Ullrich, B. Photoluminescence from two-phase nanocomposites embedded in polymers. *Micromachine* **2024**, *15*, 111. [CrossRef] [PubMed]
13. Hu, S.; Yan, X.; Zhang, Y.; Yang, B.; Li, H.; Sheng, C. Light-Induced Photoluminescence Quenching and Degradation in Quasi 2D Perovskites Film of $(C_6H_5C_2H_4NH_3)_2$ $(CH_3NH_3)_2[Pb_3I_{10}]$. *Appl. Sci.* **2021**, *11*, 2683. [CrossRef]
14. Colvin, V.L.; Schlamp, M.C.; Alivisatos, A.P. Light-emitting diodes made from cadmium selenide nanocrystals and a semiconducting polymer. *Nature* **1994**, *370*, 354–357. [CrossRef]
15. Seitz, M.; Gant, P.; Castellanos-Gomez, A.; Prins, F. Long-Term Stabilization of Two-Dimensional Perovskites by Encapsulation with Hexagonal Boron Nitride. *Nanomaterials* **2019**, *9*, 1120. [CrossRef] [PubMed]
16. Khadka, D.B.; Shirai, Y.; Yanagida, M.; Miyano, K. Degradation of encapsulated perovskite solar cells driven by deep trap states and interfacial deterioration. *J. Mater. Chem. C* **2018**, *6*, 162–170. [CrossRef]
17. Li, R.; Yi, C.; Ge, R.; Zou, W.; Cheng, L.; Wang, N.; Wang, J.; Huang, W. Room-temperature electroluminescence from two-dimensional lead halide perovskites. *Appl. Phys. Lett.* **2016**, *109*, 151101. [CrossRef]
18. Yang, Y.; Yang, M.; Zhang, S.; Lin, X.; Dang, Z.; Wang, D. Photoinduced healing of polyolefin dielectrics enabled by surface plasmon resonance of gold nanoparticles. *J. Appl. Polym. Sci.* **2019**, *10*, 47158. [CrossRef]
19. Hou, X.; Kang, J.; Qin, H.; Chen, X.; Ma, J.; Zhou, J.; Chen, L.; Wang, L.; Wang, L.W.; Peng, X. Engineering Auger recombination in colloidal quantum dots via dielectric screening. *Nat. Commun.* **2019**, *10*, 1750. [CrossRef] [PubMed]
20. Kagan, C.R.; Murray, C.B.; Nirmal, M.; Bawendi, M.G. Electronic Energy Transfer in CdSe Quantum Dot Solids. *Phys. Rev. Lett.* **1996**, *76*, 1517. [CrossRef] [PubMed]
21. Elumalai, D.; Rodríguez, B.; Kovtun, G.; Hidalgo, P.; Méndez, B.; Kaleemulla, S.; Joshi, G.M.; Cuberes, M.T. Nanostructural Characterization of Luminescent Polyvinyl Alcohol/Graphene Quantum Dots Nanocomposite Films. *Nanomaterials* **2024**, *14*, 5. [CrossRef] [PubMed]
22. Baburin, A.S.; Moskalev, D.O.; Lotkov, E.S.; Sorokina, O.S.; Baklykov, D.A.; Avdeev, S.S.; Buzaverov, K.A.; Yankovskii, G.M.; Baryshev, A.V.; Ryzhikov, I.A.; et al. Evolutionary selection growth of silver films for low-loss nanophotonic devices. *Surf. Interfaces* **2023**, *39*, 102897. [CrossRef]

coatings

Article

Future Parabolic Trough Collector Absorber Coating Development and Service Lifetime Estimation

Ana Drinčić, Luka Noč, Franci Merzel and Ivan Jerman *

National Institute of Chemistry, Hajdrihova 19, 1000 Ljubljana, Slovenia; ana.drincic@ki.si (A.D.); luka.noc@gmail.com (L.N.); franci.merzel@ki.si (F.M.)
* Correspondence: ivan.jerman@ki.si; Tel.: +386-1-476-0-440

Abstract: This work presents a study on the optical and mechanical degradation of parabolic trough collector absorber coatings produced through the spray coating application technique of in-house developed paint. The main aim of this investigation is to prepare, cure, load, and analyze the absorber coating on the substrate under conditions that mimic the on-field thermal properties. This research incorporates predicted isothermal and cyclic loads for parabolic trough systems as stresses. Biweekly inspections of loaded, identical samples monitored the degradation process. We further used the cascade of data from optical, oxide-thickening, crack length, and pull-off force measurements in mathematical modelling to predict the service life of the parabolic trough collector. The results collected and used in modelling suggested that cyclic load in combination with isothermal load is responsible for coating fatigue, influencing the solar absorber optical values and resulting in lower energy transformation efficiency. Finally, easy-to-apply coatings made out of spinel-structured black pigment and durable binder could serve as a low-cost absorber coating replacement for a new generation of parabolic trough collectors, making it possible to harvest solar energy to provide medium-temperature heat to decarbonize future food, tobacco, and paint production industrial processes.

Keywords: parabolic trough collectors; absorber coatings; service lifetime

Citation: Drinčić, A.; Noč, L.; Merzel, F.; Jerman, I. Future Parabolic Trough Collector Absorber Coating Development and Service Lifetime Estimation. *Coatings* **2024**, *14*, 1111. https://doi.org/10.3390/coatings14091111

Academic Editor: Xu Long

Received: 19 July 2024
Revised: 1 August 2024
Accepted: 14 August 2024
Published: 2 September 2024

1. Introduction

Globally, heat represents around 74 percent of the energy used by the industrial sector and contributes to over 20 percent of the total global energy consumption across all sectors [1]. Currently, the majority of industrial heat generation is derived from the process of burning fossil fuels. The most energy-intensive industrial processes, such as cement, iron and steel, and glass production, require high temperatures (>1000 °C) of industrial heat. These processes are extremely difficult to decarbonize [2]. A quite similar amount of final energy consumption to iron and steel production is present in the food, tobacco, and paper industries, but the industrial heat temperature here is much lower. In the food and tobacco industry, the majority (55%) of the industrial heat is below 100 °C and 45% is between 100 and 400 °C [3]. Decarbonization in all sectors is essential and urgent [4]. Using solar thermal energy can significantly reduce energy costs and carbon dioxide emissions for a wide range of industrial product processing applications. In the EU, the cumulated surfaces of solar thermal in operation in the European Union in 2022 are evaluated at 58.8 million m^2. In 2022, the EU concentrated solar power (CSP) capacity was assessed at 2333 MW [5]. CSP systems use parabolic trough collectors (PTCs) and linear Fresnel reflectors (LFRs) or tower technology to provide heating and cooling to manufacturing facilities. Different technologies, i.e., high-temperature heat pumps and heat storage, are combined to form a single system that can reliably provide the necessary heat for a wide variety of manufacturing processes, including those in the textile, plastics, wood, metal, and chemical industries [6].

Solar absorber coating is a crucial part of the absorber as it serves as an energy entry point, tailoring energy conversion efficiency [7]. From the perspective of the initial energy harvesting point and after a certain period of time on the field, energy transformation efficiency is important to maximize the absorber output. To maximize the absorber output, different coating types were developed in the past [8]. Coating complexity and the coating application process is consequently related to the coating's final price. Electroplating and physical vapour deposition (PVD) offer high figure of merit [9] but also higher production price. For mass production on flat surfaces, coil-coating application is suitable [10]. For pipes that are used in CSP PVD production, it is an established practice that needs high vacuum [11,12]. Thickness-sensitive spectrally selective coatings made by spray application could meet the demands of CSP [13,14]. Furthermore, the absorber coating must have high solar absorptance and low thermal emittance at the CSP operating temperature, so its initial properties and durability are critical [15,16]. Every day, the collector-coating system experiences a wide range of temperatures and solar concentration fluctuations caused by day and night exchange and cloud passage, resulting in cycling from field to maximum temperature. If the coating or the substrate is not stable enough, optimal energy transformation efficiency will fail. A decrease in energy transformation efficiency as an increase in thermal emittance and decrease in solar absorptance result from the coating deteriorating under the effects of thermal load, oxygen influence, and environmental conditions (such as dew and humidity if the absorber pipe is not in a vacuum). Coating degradation could result in optical or mechanical failure as a result of accelerated ageing that simulates the on-field situation. Estimating the lifetime of an absorber coating has historically made use of a number of different approaches. Accelerated ageing was performed through extensive thermal cycling [17], isothermally aged in air [18,19], under real high solar flux [20] and thermal cycling [15,21,22]. Research often solely focuses on studying fatigue materials and their consequences. Rarely, extensive research is conducted that includes service-life predictions. A procedure for accelerated life testing of flat plate solar absorber surfaces was developed within the framework of the Materials in Solar Thermal Collectors working group of the International Energy Agency--Solar Heating and Cooling Programme [23,24]. Out of Task X, a European standard (EN 12975-3-1 [25]) and international standards (ISO 22975-3:2014 [26]) were made. In our opinion, such an approach for PTC absorber coatings and tower coatings is delicate, as in Task X for the estimation of activation energy, where they used much higher temperatures than the operation temperature. This is not a problem when you are testing vented flat plate solar absorber coatings for collectors working under conditions corresponding to those in a typical solar domestic hot water system or combi system. In the aforementioned case, a higher temperature than the stagnation temperature used in accelerated testing does not cause fatigue to the substrate (copper or aluminium). The oxidation of the polished surface beneath the coating, which accounts for the increase in thermal emittance value, is negligible. The fact that aluminium thickness stops at a few nm in air at temperatures around 100 °C supports that statement [27]. In the case of parabolic trough collectors, the operation temperature is much higher in contrast to a vented flat plate collector as a result of the higher concentration of solar energy on the small surface. Consequently, operation temperature is often already limiting material use at high pressure, and the additional step towards higher temperature needed for achieving the acceleration factor is not acceptable. Nevertheless, the findings of this study broaden the range of the model put forward in our previous research [15] in the direction of parabolic trough collectors. The objective of this study was to establish the protocol that determines the lifetime of scalable parabolic trough collector coatings made by, i.e., electrodeposition, deep coating, and spray coating applications. The coatings undertook laboratory testing to simulate the loads that could be experienced on-field, demonstrating their fatigue behaviour while maintaining excellent optical characteristics. The theoretical model integrated both mechanical and optical degradation by using variables such as solar absorptance, thermal emittance, oxide thickness, and fracture evaluation values to accurately predict the service lifetime of the coating. We present a laboratory-based degradation evaluation through visual/infrared and X-ray diffraction (XRD) spectroscopy, scanning electron microscopy (SEM), and energy-dispersive X-ray spectroscopy (EDS). Our investigation concentrates on assessing the timeline of energy transformation efficiency decline for industrial coatings. We analyse both initial and loaded coatings to provide comprehensive data. The approach given here can be utilised to evaluate coatings utilised for competitive solar heat feeding in future Solar Heat Industrial Processes (SHIPs). We show degradation information from a lab point of view using X-ray diffraction (XRD) spectroscopy, scanning electron microscopy (SEM), and energy-dispersive X-ray spectroscopy (EDS) on both the initial and loaded coatings. This helps us figure out how long it takes for industrial coatings to lose their ability to efficiently transform solar energy into useful heat. Coatings used for competitive solar heat feeding the future industrial processes using solar heat can be evaluated using the approach provided here. It is important to emphasise that in order to obtain an accurate result, we must assess the entire system, including the coating and substrate.

2. Materials and Methods

2.1. Coating Preparation

We added the pigment (Black 444, Shepherd Colour Company, West Chester Township, OH, USA) to a suitable amount of resin (Dowsil 2405, Dow, Torrance, CA, USA), a solvent (xylene, Fluka), and additives. We then stirred the mixture with the dissolver VMA Dispermat CN10 for 30 min at 1200 rpm. Furthermore, we milled the pigment dispersions in a Dyno-mill Research Laboratory agitator bead mill (WAB, Willy A. Bachofen AG Maschinenfabrik, Muttenz, Switzerland) using 0.3 mm diameter zirconia beads for half an hour at 5000 rpm. We diluted the nanoparticle pigment dispersion to the appropriate viscosity after milling and then prepared it for the spray coating. Coupons made out of steel 5×5 cm in size were polished by grid paper (1000). After spray application, wet coating thickness was 40 microns. Coatings were cured in an air oven according to established protocol [16].

2.2. Coating Load

Using a series of computer-controlled furnaces equipped with humidity/dew generators and increased oxygen concentration settings, we simulated the thermal load normally faced by solar loop absorbers supplying industrial heat in a lab setting. Following the coating curing protocol, Figure 1 schematically presents the loading protocol for identical samples of PTC absorber coatings. We performed iso-thermal loading at 300, 400, and 500 °C. We exposed the samples in parallel to a cyclic load consisting of four steps. The first step involved simulating dew and purging the furnace with steam. After 5 min of dew simulation at room temperature, exponential heating takes place, followed by a short plateau phase (300, 400, and 500 °C) and exponential cooling to room temperature. Heating and cooling temperature was set to 6, 8, and 10 °C/min. In addition to air atmosphere in furnaces, in one example oxygen was added to the furnace.

2.3. Characterization

We assessed weekly surface crack propagation using optical microscopy, evaluated the weekly growth of the oxide layer beneath the coating using metallurgical cross-cut evaluation and a Zeiss UltraPlus (Jena, Germany) analytical low-voltage scanning electron microscope (SEM), estimated adhesion using pull-off testing, and gained a comprehensive understanding of the collector material, the growth of the oxide layer below, and the diffusion processes inside the collector surface using EDS. To obtain X-ray powder diffraction (XRD) patterns, an X-ray powder diffractometer (PANalytical X'Pert PRO, Malvern, UK) with CuKα radiation (λ = 1.5406 Å) was used. The 2θ angle was set between 10° and 70°, and the scanning speed was 0.04°/s. A Lambda 950 UV/VIS spectrometer was used to measure the near normal spectral reflectance of the samples at wavelengths from 250 nm to 2500 nm. We used a Bruker Vertex 70v spectrometer (Billerica, MA, USA) with gold-coated integrating sphere equipment to measure the samples' normal hemispherical directional reflectance spectra in the infrared range, which covers wavelengths from 1.5 to 16.5 m. We obtained the values for solar absorptance (α_S; Equation (1)) and thermal emittance (ε_T; Equation (2)) from the measured reflectance spectra calculated numerically from the measured reflectance spectra at room temperature using the equations below:

$$\alpha_S = \frac{\int_{0.25\ \mu m}^{2.5\ \mu m} I_s(\lambda)(1 - R(\lambda))d\lambda}{\int_{0.25\ \mu m}^{2.5\ \mu m} I_s(\lambda)d\lambda} \quad (1)$$

$$\varepsilon_T = \frac{\int_{1.5\ \mu m}^{16.5\ \mu m} I_b(\lambda)(1 - R(\lambda))d\lambda}{\int_{1.5\ \mu m}^{16.5\ \mu m} I_b(\lambda)d\lambda} \quad (2)$$

where $I_s(\lambda)$ is the reference solar spectral irradiance of AM 1.5 according to the ISO standard (ISO 9845-1, 1992 [28]), $R(\lambda)$ is the spectral reflectance, and I_b (λ) is the blackbody radiation at 80 °C.

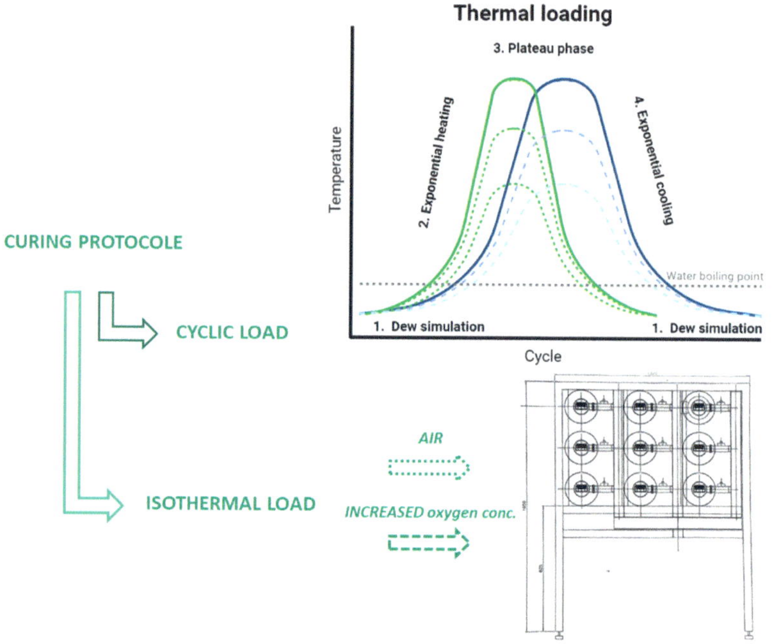

Figure 1. Schematic presentation of thermal load that simulates the on-field stress.

2.4. Mathematical Modelling

In order to obtain a quantitative description of the degradation of the PTC absorber coatings, we use a mathematical model which we proposed in our previous study (15), which is designed to explicitly take into account the isothermal- and thermal-cycling loads to which the absorber coating is exposed during its operation. The model distinguishes between the thermal cycles performed in the dry and humid atmosphere. It is assumed in the model that the optical degradation processes in the PTC absorber coating can be described by the first-order kinetic equation:

$$\frac{d\alpha}{dt} = k(T,\ \Delta T)f(\alpha) \quad (3)$$

where α is the extent of the conversion of the observed quantity (i.e., absorptance), $f(\alpha)$ is the relevant reaction model and $k(T,\Delta T)$ is the rate constant, which depends on the temperature (T) during the isothermal load (it) and the magnitude of the temperature difference ΔT during dry (dc) or wet/humid (wc) thermal cycling. Here, the degradation due to the isothermal load is given as a function of time, while the thermal cycling is given as a function of the number of cycles (n). For the model evaluation, it should be known in advance how many thermal cycles of each type (dry/wet) are expected during a given time interval. As the loads are not present simultaneously, we can write to a good approximation a cumulative rate of the degradation as a sum of individual contributions:

$$\frac{d\alpha}{dt} = \left(\frac{d\alpha}{dt}\right)_{it} + \tau_{dc}\left(\frac{d\alpha}{dn}\right)_{dc} + \tau_{wc}\left(\frac{d\alpha}{dn}\right)_{wc} \tag{4}$$

Each term is supposed to have its own specific form. In particular, the isothermal rate constant is assumed to be of the Arrhenius type [24], and the cycling rate constant is assumed to have a power-law dependence on the temperature change, ΔT, according to the Coffin–Manson relationship [29], which is commonly applied in modelling the fatigue failure of materials subjected to thermal cycling. The factors $\tau_{dc/wc} = dn/dt$ are added in order to keep explicit time dependence in the equation.

The individual reaction models $f_i(\alpha)$ (i = {1 = it, 2 = dc, 3 = wc}) differ from each other, as well as model parameters A_i, e_0, and β_i, which are determined in the separate sets of experiments dealing exclusively with one type of the degradation load. Fitting procedures are used for deriving individual parameters.

$$\left(\frac{d\alpha}{dt}\right)_{it} = k_1(T)f_1(\alpha) = A_1 exp\left(-\frac{e_0}{T}\right)f_1(\alpha) \tag{5}$$

$$\left(\frac{d\alpha}{dt}\right)_{dc} = k_2(\Delta T)f_2(\alpha) = A_2 exp\left(-\frac{e_0}{T}\right)^{\beta_2} f_2(\alpha) \tag{6}$$

$$\left(\frac{d\alpha}{dt}\right)_{wc} = k_3(\Delta T)f_3(\alpha) = A_3 exp\left(-\frac{e_0}{T}\right)^{\beta_3} f_3(\alpha) \tag{7}$$

3. Results

3.1. Crack Evolution

Crack evolution can affect the coating's optical properties by trapping part of the solar energy inside the cracks, thereby increasing the absorptance if cracks and pores are on a proper micro to nano level [30]. Conversely, if the cracks are too large, the substrate will affect the absorptance values, leading to decreased absorptance. We evaluated the cracks in our samples on identical location every two weeks, using an optical microscope and cured them before beginning testing. After each measurement, we processed the images with custom-developed software to estimate the length and thickness of any cracks present in the coating. We should highlight that in the case of PTC coatings made by spray-out of pigment dispersion prepared as described above and loaded at temperatures 300, 400, and 500 °C after proper curing, we did not observe the paint shrinkage that resulted in cracks on the majority of samples if we applied proper coating thickness. Figure 2 represent examples at 300 °C, where the coating thickness limits the optimal thickness. During the first heat-exposure flux, the formation of minor cracks was detected on the surface of the coating as a result of the difference in the thermal expansion of the substrate, the binder decomposition and shrinkage, the formation of the oxide layer, and the coating densification. The appearance of cracks was only spotted in two of our samples, but we did not notice growth as a function of loading time.

On other samples, we did not discover cracks as the samples were stable enough at the tested temperature.

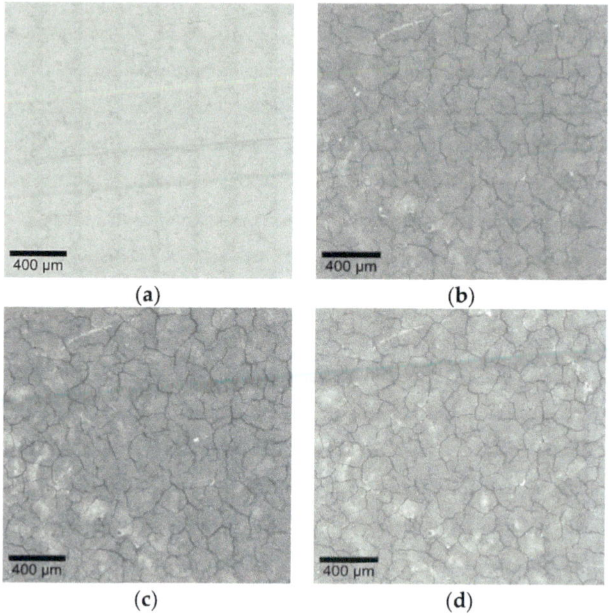

Figure 2. Optical microscopy images for pigmented coatings as a function of time under thermal load at 300 °C after (**a**) initial application; (**b**) 2 weeks; (**c**) 4 weeks; (**d**) 6 weeks.

3.2. Oxide Thickness Evaluation

We used metallurgical cross-cutting, along with SEM microscopy, to study and understand the interface between the substrate and coating. According to Figure 3, the coating changes with load and stress. The picture shows how the oxide protective layer connects to the coating layer that forms on the substrate surface. Moreover, oxide thickness is estimated on 653 nm after 591 h of iso-thermal load at 400 °C.

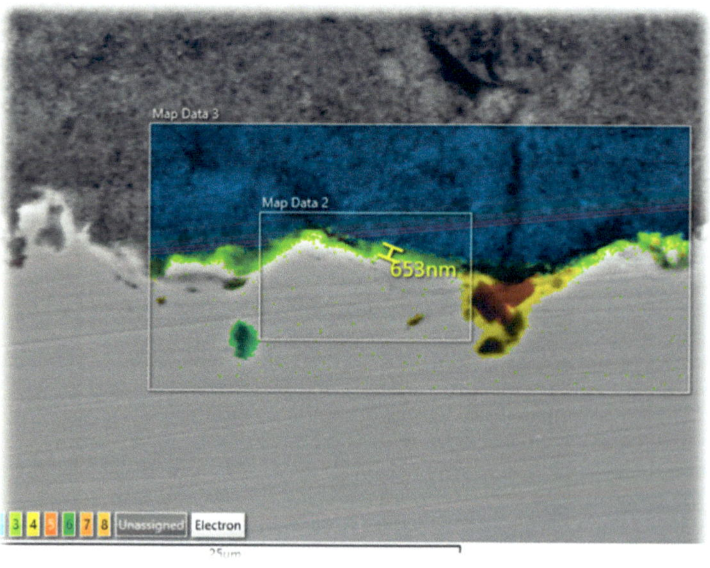

Figure 3. Cross-cut of coating SEM/EDS image after 591 h at 400 °C.

The growth of the oxide layer during thermal loading is significantly slower at lower temperatures, whereas it grows more quickly at higher temperatures (Figure 4). In the case of cyclic thermal loading, the formation of the oxide layer was similar to that detected at high temperatures since the cyclic loading also involved exposure to 500 °C in the plateau phase. Furthermore, it is also evident that exposure to a lower (300 and 400 °C) load temperature causes significantly slower oxide thickness growth. Despite over 2000 h at these temperatures, the oxide thickness stays below 1 micron, suggesting that the dense oxide layer does not limit or weaken adhesion, even after prolonged accelerated stress.

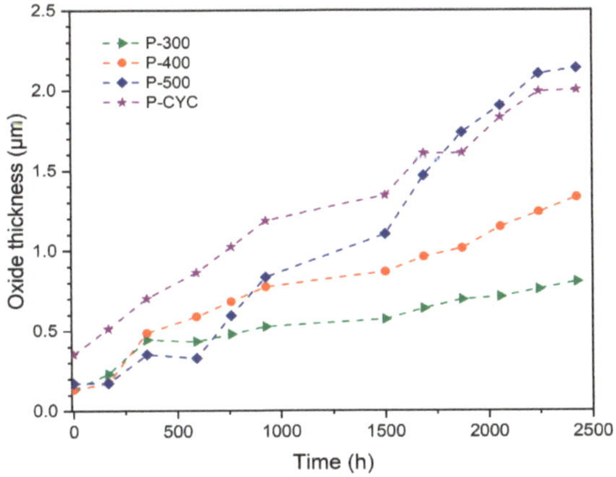

Figure 4. Oxide thickness as a function of time for different isothermal and cyclic loads.

3.3. Adhesion Evaluation

The oxygen ion diffusion distance and load temperature directly influence the growth rate of thermally grown oxide on the substrate. The spray-deposited absorber coatings are a consequence of solvent-borne paint being porous after curing, so oxygen has limited access to the substrate surface. If the oxide layer's non-uniform thickness is too high, it may change the local stress state and cause the absorber coating to crack. As a result of crack evolution and oxide growth, coatings can de-bond after a thermal load. The coating layer peeling off from the substrate or cracking within a coating layer can lead to the failure of the entire optical coating system. The latter is more prevalent at temperatures above 750 °C [31]. To evaluate the mechanical failure in the coating, we studied adhesion strength as a function of time during thermal load exposure. We measured the release behaviour using a standard pull-off test, which assesses the adhesive properties by measuring the stress required to pull away the dolly from a flat, rigid, cylindrical punch initially pressed into contact with the adhesive surface. Our tests provided a time evolution of the average critical pull-off stress for different loads. By inspecting the punch surface, we identified different delamination layers. Figure 5 represents the dolly surface after the pull-off test at the different temperatures. When epoxy glue adheres the dolly to the loaded paint surface, it delaminates at its weakest point. If the adhesion is good, the weakest point will be the glue, which appears as a white dolly surface. When paint decomposes or lacks sufficient binder, it becomes the weakest point, and its black pigment composition results in a black dolly surface. Furthermore, if thick oxide grows below the surface, delamination occurs over the oxide layer. In the last scenario, the dolly surface should be grey. As shown in Figure 5, the most common colours on dolly surfaces are white and black. This indicates excellent paint adhesion with some delamination. That is why additional inspections are necessary if delamination takes place in the paint layer. Most often, we witness delamination with a very thin layer. As the surface is very rough, we remove only the top layer.

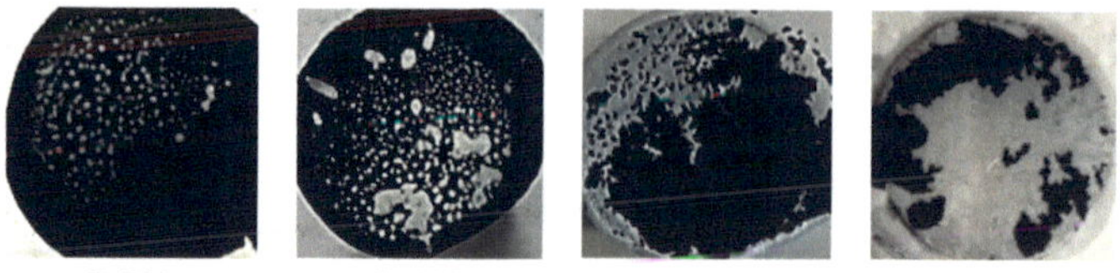

Figure 5. Dolly surface after pull-off test photos for samples after 1314 h under different thermal loads.

Furthermore, we measured the force needed to remove the glued dolly from the substrate. Figure 6 presents the results, demonstrating the stable force required to remove the dolly over time for samples loaded at 300 and 400 °C. The pull-off test's value for dolly debonding was initially around 13 MPa and displayed rather steady behaviour, with a slightly increasing trend over time. For samples loaded under cyclic load or at 500 °C at the beginning, for the first 1000 h, the pull-off force remains constant, but after 1000 h of load, it looks like the oxide thickness reaches a limiting thickness and the pull-off force decreases with time [15]. After this period, we noticed a discrepancy in adhesion force for samples at higher temperatures. Different modes of delamination can explain the variation in the pull-off stress values, as illustrated in Figure 5.

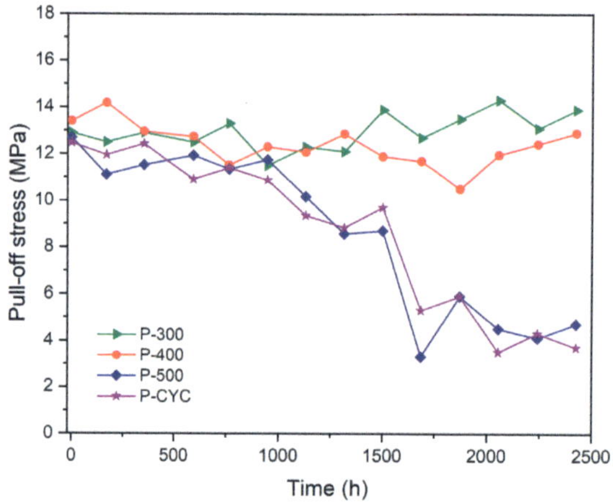

Figure 6. Pull-off force as a function of loading time.

3.4. *Coating Optical Property Changes Due to Thermal Load*

In general, with some exceptions, researchers develop, classify, and select absorber coatings based solely on their peak thermal performance, disregarding factors such as cost, durability, and operating conditions. In our opinion, durability and cost are crucial when we are talking about mass production. The optical properties have a direct impact on the levelized energy cost. As a result, it is necessary to measure the spectral absorptance and thermal emittance of new and stressed coatings. A perfect spectrally selective absorber has almost no reflectance at short wavelengths and a lot of reflectance at long wavelengths. The operating temperature determines the sharp change between these two regimes (Figure 7). However, the overall properties across the solar spectral range are significant, as thermal emittance must also be considered, particularly for spectrally selective coatings. Substrate polish and the mirror-like structure are impacted by thermal emittance. Applying coatings to a sand-blasted surface results in a thermal emittance of approximately 80%. Furthermore, decreasing coating thickness directly affects thermal emittance because the IR penetration range is limited. We should keep in mind that too low a thickness means non-maximal solar absorptance; therefore, there is always a trade-off. For the coating used in our experiments, we measured a very high absorptance (>0.96), but not low emittance, because they are non-selective. Also, we discovered that the absorptance does not depend on the thickness (Figure 8). We prepared wet 5, 10, 20 and 40 μm thick coatings and measured absorption and emittance. As you can see from the figure, we achieved similar absorptance of 96% plus, but unfortunately, we have an emittance over 80% if the substrate is not polished.

Wet film thickness:	5 μm	10 μm	20 μm	40 μm
α_S value:	0.962	0.962	0.963	0.963
ε_T value:	0.800	0.830	0.859	0.867

Figure 7. Dependence of coating thickness on solar absorptance and thermal emittance values.

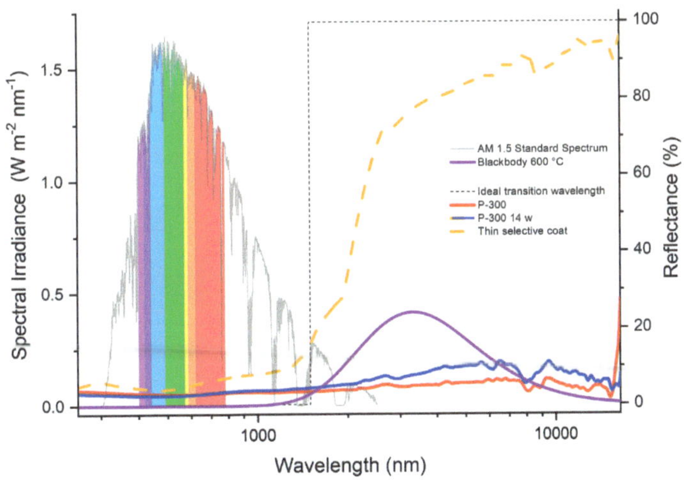

Figure 8. Reflectance spectra of coatings with B444 pigment before thermal ageing and after 14 weeks (red and blue) at 300 °C. The orange spectrum represents pigmented coating on polished surface, which is close to the ideal spectrally selective coating (dashed line). In addition, graphs for blackbody radiation at 600 °C (violet) and AM 1.5 reference solar spectra (grey) are shown.

To achieve lower emittance, a coating with around 1.5 μm thickness needs to be prepared on a polished substrate. Figure 8 represents the reflection spectra for the visual and IR ranges. In orange, a thin selective coating is presented on the surface of a polished substrate. That shape is similar to that of the dashed line, indicating an almost ideal selective coating. A delta ε_TT of 60% relates to the difference in spectra between P-300 and thin selective coatings.

Furthermore, every two weeks, we measured how solar absorptance changed with isothermal and cyclic thermal loads. Figure 9 displays that we achieved more than 96% solar absorptance for our samples. We also discovered that the coatings showed different degradation at different temperatures, but we want to emphasize that after 2500 h, the developed coatings' solar absorptance did not drop below the threshold value. A slight increase in solar absorptance is seen at the beginning, and later values drop between 1000 and 1500 h of thermal load. In our opinion, this shift is related to measurement error and averaging the sample values as the deviation is in the range of 0.5%. These observations are consistent with the variation in their solar absorptance during ageing. Note that the upper and lower error bar values indicate the maximum and minimum measured solar absorptance values from the batch of samples aged under the same conditions. If we performed proper curing, the solar absorptance of our coatings remained relatively stable. However, absorptance continued decreasing only if the loading temperature was above 700 °C (not part of the article).

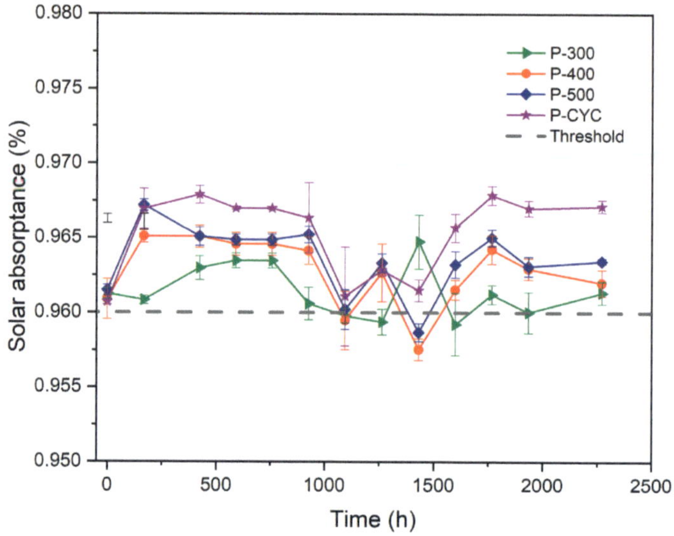

Figure 9. Solar absorptance as a function of time after the isothermal thermal load for coatings made out of pigments.

3.5. *XRD Spectroscopy Evaluation*

We used X-ray diffraction (XRD) to look at the crystal structure of the pigment and the oxide growth of the substrate-protecting layer to find the weakest spot in the solar absorber's coating. XRD diffractograms of the initial substrate and coated samples (Figure 10) after one week of isothermal testing at 300, 400, and 500 °C in an air furnace are presented. Examining the substrates reveals that a thin oxide layer is present after one week of exposure to 500 °C. We observed the martensitic and austenitic phases in untreated specimens. A mechanical polishing effect is associated with the martensitic presence. The thermal load intensifies the emphasis on austenitic phases. Over time, we also note the formation of Cr_2O_3 [32]. Spinel pigment is very stable and does not show any change in crystal structure at temperatures up to 500 °C. At higher temperatures, we witness changes related to diffusion [16].

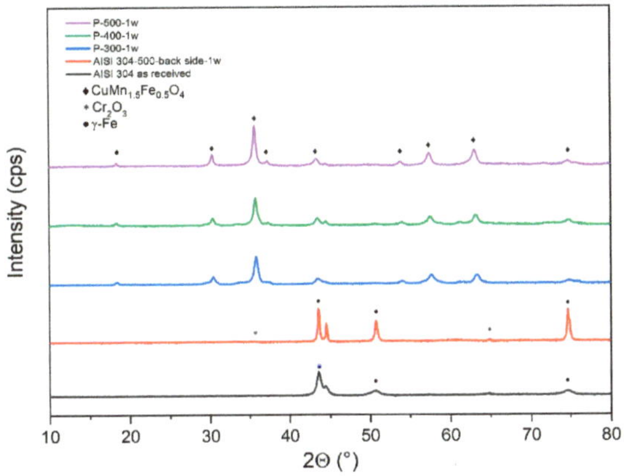

Figure 10. XRD spectra of the substrate and coated samples after one week of isothermal testing at 300, 400, and 500 °C in air furnace.

3.6. Service Lifetime Prediction

In the final step, we used a mathematical model represented in Equation (2) for calculating optical degradation as a function of time. We fitted the experimental data for each temperature and cycling type to model the degradation curve. From the degradation curve, we infer a lifetime of 15 years (the intersection between the black and the threshold curves) when both loads are taken into account. After 15 years, we can expect that solar absorptance will begin to decrease below 0.96. The prediction has a 16% relative error (Figure 11). Over a span of 15 years, if the suggested coating is used to harness solar energy for industrial processes, there is potential for a significant reduction in CO_2 emissions by the industry.

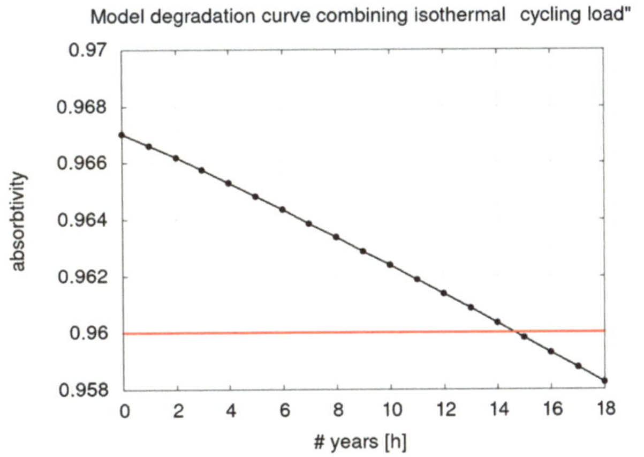

Figure 11. Model degradation curve for solar absorptance as a function of time on the field (same load); threshold value is marked with red line.

4. Conclusions

Our research provides valuable insights into the preparation and evaluation of the PTC coating lifetime, with the aim of developing competitive coatings that could be applied via the spray method. We extensively studied coatings not only for their longevity and lifetime assessment but also to gather information for improving the next generation of PTC absorber coatings. Selecting the right coatings for the desired substrate and targeting the application and curing of the PTC coating typically results in excellent longevity.

To accurately predict the lifetime, we combined isothermal and cyclic loads. We developed a mathematical model that includes all the needed aspects for accurate lifetime prediction. We also looked at optical and mechanical degradation separately to determine which of the two progresses more rapidly towards the threshold value. The more progressive degradation means that the coating needs to be reapplied sooner.

Future research should investigate low-cost absorber coatings for PTC technology that offer reliable heat harvesting for the future decarbonised food, paper, and textile industries. However, it is important to approach price reductions with caution, as undue undercutting can damage a brand's reputation and profitability.

Author Contributions: A.D., L.N. and I.J. conceived and designed the research, performed data analysis, and wrote the manuscript. L.N. performed material synthesis, application, and characterization. F.M. conducted the mathematical model and wrote and reviewed the manuscript. All authors discussed the results and commented on the manuscript. The manuscript was written through the contributions of all authors. All authors have read and agreed to the published version of the manuscript.

Funding: This research was funded by EU Horizon 2020 (grant no. 884213) and Agency for Research and Innovation Slovenia (grant no. P2-0393).

Data Availability Statement: Data are contained within the article.

Acknowledgments: The authors acknowledge the research core funding No. P2-0393 and the project Forthcoming Research and Industry for European and National Development of SHIP (FRIENDSHIP), No 884213). The authors thank Edi Kranjc (National Institute of Chemistry Ljubljana) for XRD measurements. All authors discussed the results and commented on the manuscript.

Conflicts of Interest: The authors declare no conflicts of interest.

References

1. Thiel, G.P.; Stark, A.K. To Decarbonize Industry, We Must Decarbonize Heat. *Joule* **2021**, *5*, 531–550. [CrossRef]
2. Davis, S.J.; Lewis, N.S.; Shaner, M.; Aggarwal, S.; Arent, D.; Azevedo, I.L.; Benson, S.M.; Bradley, T.; Brouwer, J.; Chiang, Y.-M.; et al. Net-Zero Emissions Energy Systems. *Science* **2018**, *360*, eaas9793. [CrossRef] [PubMed]
3. Naegler, T.; Simon, S.; Klein, M.; Gils, H.C. Quantification of the European Industrial Heat Demand by Branch and Temperature Level. *Int. J. Energy Res.* **2015**, *39*, 2019–2030. [CrossRef]
4. European Commission; Directorate-General for Communication. *Decarbonising Our Energy System to Meet Our Climate Goals*; Publications Office of the European Union: Luxembourg, 2021. [CrossRef]
5. EurObserv'ER. *Solar Thermal and Concentrated Solar Power Barometer 2023*; EurObserv'ER: Paris, France. Available online: https://www.eurobserv-er.org/solar-thermal-and-concentrated-solar-power-barometer-2023/ (accessed on 17 July 2024).
6. Kabir, E.; Kumar, P.; Kumar, S.; Adelodun, A.A.; Kim, K.-H. Solar Energy: Potential and Future Prospects. *Renew. Sustain. Energy Rev.* **2018**, *82*, 894–900. [CrossRef]
7. Noc, L.; Jerman, I. Review of the Spectrally Selective (CSP) Absorber Coatings, Suitable for Use in SHIP. *Sol. Energy Mater. Sol. Cells* **2022**, *238*, 111625. [CrossRef]
8. Cao, F.; McEnaney, K.; Chen, G.; Ren, Z. A Review of Cermet-Based Spectrally Selective Solar Absorbers. *Energy Environ. Sci.* **2014**, *7*, 1615–1627. [CrossRef]
9. Roberts, D.E. A Figure of Merit for Selective Absorbers in Flat Plate Solar Water Heaters. *Sol. Energy* **2013**, *98*, 503–510. [CrossRef]
10. Kozelj, M.; Orel, B.; Steinbuecher, M.; Jerman, I.; Vodlan, M. Aminosilane Modified Pigments for Spectrally Selective Coatings, Method for Preparation and Their Use in Coatings. EP2261288B1, 16 January 2019.
11. Kuckelkorn, T.D.; Silmy, K.D.; Dreyer, S.D. Radiation-Selective Absorber Coating and Absorber Tube with Radiation-Selective Absorber Coating. EP2312234A2, 20 April 2011.
12. Asaad Yasseen, A.-R.; Istvan, S.; Istvan, F. Selective Absorber Coatings and Technological Advancements in Performance Enhancement for Parabolic Trough Solar Collector. *J. Therm. Sci.* **2022**, *31*, 1990–2008. [CrossRef]
13. Jerman, I.; Zepeda, F.; Noc, L. Merging of Oxide Species with Black Spinel Structure by CSP Operating Temperature. *AIP Conf. Proc.* **2018**, *2033*, 220002. [CrossRef]
14. Jerman, I.; Kovac, J.; Colovic, M.; Zepeda, F.; Noc, L. POSS-Modified Black Pigment for CSP Deployment. *AIP Conf. Proc.* **2018**, *2033*, 220003. [CrossRef]
15. Noc, L.; Sest, E.; Kapun, G.; Ruiz-Zepeda, F.; Binyamin, Y.; Merzel, F.; Jerman, I. High-Solar-Absorptance CSP Coating Characterization and Reliability Testing with Isothermal and Cyclic Loads for Service-Life Prediction. *Energy Environ. Sci.* **2019**, *12*, 1679–1694. [CrossRef]

16. Noc, L.; Ruiz-Zepeda, F.; Merzel, F.; Jerman, I. High-Temperature "Ion Baseball" for Enhancing Concentrated Solar Power Efficiency. *Sol. Energy Mater. Sol. Cells* **2019**, *200*, 109974. [CrossRef]
17. Hosseini, S.; Torres, J.F.; Taheri, M.; Tsuda, K.; Tricoli, A.; Lipiński, W.; Coventry, J. Aging Mechanisms for High-Temperature Solar Absorber Coatings under Extensive Thermal Cycling. *Sol. Energy Mater. Sol. Cells* **2024**, *271*, 112856. [CrossRef]
18. Ambrosini, A.; Lambert, T.N.; Boubault, A.; Hunt, A.; Davis, D.J.; Adams, D.; Hall, A.C. Thermal Stability of Oxide-Based Solar Selective Coatings for CSP Central Receivers; American Society of Mechanical Engineers Digital Collection. *Energy Sustain.* **2015**, *56840*, V001T05A022. [CrossRef]
19. Wang, X.J.; Jiang, L.J.; Du, M.; Hao, L.; Liu, X.P.; Yu, Q.H. The Degradation of Solar Thermal Absorption Coatings. *Energy Procedia* **2014**, *49*, 1747–1755. [CrossRef]
20. Reoyo-Prats, R.; Carling Plaza, A.; Faugeroux, O.; Claudet, B.; Soum-Glaude, A.; Hildebrandt, C.; Binyamin, Y.; Agüero, A.; Meißner, T. Accelerated Aging of Absorber Coatings for CSP Receivers under Real High Solar Flux—Evolution of Their Optical Properties. *Sol. Energy Mater. Sol. Cells* **2019**, *193*, 92–100. [CrossRef]
21. Torres, J.F.; Ellis, I.; Coventry, J. Degradation Mechanisms and Non-Linear Thermal Cycling Effects in a High-Temperature Light-Absorber Coating. *Sol. Energy Mater. Sol. Cells* **2020**, *218*, 110719. [CrossRef]
22. Martínez, N.; Lopez-Herraiz, M.; Rico, A.; Múnez, C.J.; Poza, P. Influence of Different Thermal Degradation Processes on the Optical Property of Pyromark-2500. *Sol. Energy* **2023**, *253*, 58–72. [CrossRef]
23. Carlsson, B.; Frei, U.; Köhl, M.; Möller, K. *Accelerated Life Testing of Solar Energy Materials—Case Study of Some Selective Solar Absorber Coatings for DHW Systems*; International Energy Agency, Solar Heating and Cooling Programme Task X: Solar Materials Research and Development, SP-Report 1994:13; SP Sveriges Tekniska Forskningsinstitut: Lund, Sweden, 1994. Available online: http://urn.kb.se/resolve?urn=urn:nbn:se:lnu:diva-3496 (accessed on 26 April 2010).
24. Brunold, S.; Frei, U.; Carlsson, B.; Möller, K.; Köhl, M. Accelerated Life Testing of Solar Absorber Coatings: Testing Procedure and Results. *Sol. Energy* **2000**, *68*, 313–323. [CrossRef]
25. *EN 12975-3-1*; Solar Collectors—General Requirements. Slovenian Institute for Standardization: Ljubljana, Slovenia, 2022.
26. *ISO 22975-3:2014*; Solar Energy—Collector Components and Materials—Part 3: Absorber Surface Durability. International Organization for Standardization: Geneva, Switzerland, 2014.
27. Cai, N.; Zhou, G.; Müller, K.; Starr, D.E. Temperature and Pressure Dependent Mott Potentials and Their Influence on Self-Limiting Oxide Film Growth. *Appl. Phys. Lett.* **2012**, *101*, 171605. [CrossRef]
28. *ISO 9845-1*; Solar Energy—Reference Solar Spectral Irradiance at the Ground at Different Receiving Conditions—Part 1: Direct Normal and Hemispherical Solar Irradiance for Air Mass 1,5. International Organization for Standardization: Geneva, Switzerland, 1992.
29. Manson, S.S. Behavior of Materials under Conditions of Thermal Stress. 1954. Available online: https://ntrs.nasa.gov/citations/19930092197 (accessed on 19 July 2024).
30. Torres, J.F.; Tsuda, K.; Murakami, Y.; Guo, Y.; Hosseini, S.; Asselineau, C.-A.; Taheri, M.; Drewes, K.; Tricoli, A.; Lipiński, W.; et al. Highly Efficient and Durable Solar Thermal Energy Harvesting via Scalable Hierarchical Coatings Inspired by Stony Corals. *Energy Environ. Sci.* **2022**, *15*, 1893–1906. [CrossRef]
31. Hosseini, S.; Torres, J.F.; Taheri, M.; Tricoli, A.; Lipiński, W.; Coventry, J. Stability and Characterisation of Pyromark 2500 Cured at Different Temperatures. *Sol. Energy* **2023**, *265*, 112102. [CrossRef]
32. Issartel, C.; Buscail, H.; Caudron, E.; Cueff, R.; Riffard, F.; Perrier, S.; Jacquet, P.; Lambertin, M. Influence of Nitridation on the Oxidation of a 304 Steel at 800 °C. *Corros. Sci.* **2004**, *46*, 2191–2201. [CrossRef]

coatings

Article

Analyses of the Properties of the NiO-Doped Ga_2O_3 Wide-Bandgap Semiconductor Thin Films

Cheng-Fu Yang [1,2], En-Chi Tsao [1], Yi-Wen Wang [1], Hsin-Pei Lin [1], Teen-Hang Meen [3,*] and Shu-Han Liao [4,*]

1 Department of Chemical and Materials Engineering, National University of Kaohsiung, Kaohsiung 811, Taiwan; cfyang@nuk.edu.tw (C.-F.Y.); a1097135@mail.nuk.edu.tw (E.-C.T.); m1125618@mail.nuk.edu.tw (Y.-W.W.); a1115647@mail.nuk.edu.tw (H.-P.L.)
2 Department of Aeronautical Engineering, Chaoyang University of Technology, Taichung 413, Taiwan
3 Department of Electronic Engineering, National Formosa University, Yunlin 632, Taiwan
4 Department of Electronic and Computer Engineering, Tamkang University, New Taipei City 251, Taiwan
* Correspondence: thmeen@gs.nfu.edu.tw (T.-H.M.); shliao@gms.tku.edu.tw (S.-H.L.)

Abstract: The study began by pre-sintering Ga_2O_3 powder at 950 °C for 1 h, followed by the preparation of a mixture of Ga_2O_3 and 12 at% NiO powders to fabricate a source target material. An electron beam (e-beam) system was then used to deposit NiO-doped Ga_2O_3 thin films on Si substrates. X-ray diffraction (XRD) analyses revealed that the pre-sintered Ga_2O_3 at 950 °C exhibited β-phase characteristics, and the deposited NiO-doped Ga_2O_3 thin films exhibited an amorphous phase. After the deposition of the NiO-doped Ga_2O_3 thin films, they were divided into two portions. One portion underwent various analyses directly, while the other was annealed at 500 °C in air before being analyzed. Field-emission scanning electron microscopy (FESEM) was utilized to process the surface observation, and the cross-sectional observation was primarily used to measure the thickness of the NiO-doped Ga_2O_3 thin films. UV-Vis spectroscopy was used to calculate the bandgap by analyzing the transmission spectra, while the Agilent B1500A was employed to measure the I-V characteristics. Hall measurements were also performed to assess the mobility, carrier concentration, and resistivity of both NiO-doped Ga_2O_3 thin films. The first innovation is that the 500 °C-annealed NiO-doped Ga_2O_3 thin films exhibited a larger bandgap and better electrical conductivity. The manuscript provides an explanation for the observed increase in the bandgap. Another important innovation is that the 500 °C-annealed NiO-doped Ga_2O_3 thin films revealed a high-energy bandgap of 4.402 eV. The third innovation is that X-ray photoelectron spectroscopy (XPS) analyses of the $Ga_{2p3/2}$, $Ga_{2p1/2}$, Ga_{3d}, $Ni_{2p3/2}$, and O_{1s} peaks were conducted to further investigate the reasons behind the enhanced electrical conductivity of the 500 °C-annealed NiO-doped Ga_2O_3 thin films.

Keywords: NiO-doped Ga_2O_3; wide-bandgap semiconductor; UV-Vis spectroscopy; electrical conductivity; X-ray photoelectron spectroscopy

Citation: Yang, C.-F.; Tsao, E.-C.; Wang, Y.-W.; Lin, H.-P.; Meen, T.-H.; Liao, S.-H. Analyses of the Properties of the NiO-Doped Ga_2O_3 Wide-Bandgap Semiconductor Thin Films. *Coatings* **2024**, *14*, 1615. https://doi.org/10.3390/coatings14121615

Academic Editor: Andrey V. Osipov

Received: 22 November 2024
Revised: 17 December 2024
Accepted: 21 December 2024
Published: 23 December 2024

1. Introduction

β-phase Ga_2O_3 is a wide-bandgap semiconductor (4.9 eV) with excellent physical and chemical stabilities that has garnered significant interest in recent years [1]. Compared with other wide-bandgap (WBG) semiconductors such as GaN and SiC, β-phase Ga_2O_3 offers faster electron drift velocities and higher breakdown voltages, and it can be produced at high volume and lower cost. Directly fabricating β-phase Ga_2O_3 thin films is not an easy task, primarily due to the material's relatively high melting point (around 1725 °C), which makes processing in high-temperature environments particularly challenging. In addition, β-phase Ga_2O_3 is an oxide material with an asymmetric crystal structure, which can lead to the formation of crystal defects or uneven growth during the thin-film deposition process, ultimately affecting the performance of the thin films. In the past, chemical vapor deposition (CVD) has been one of the primary methods for depositing Ga_2O_3 thin films. For instance, thermal CVD is particularly advantageous due to its use of environmentally friendly and cost-effective precursors [2]. The main precursors for the method of thermal CVD to deposit Ga_2O_3 thin films include gallium trichloride ($GaCl_3$), gallium nitride (GaN), gallium oxide (Ga_2O_3), and metallic gallium (Ga) [3]. While CVD-based methods are commonly used deposition techniques, using traditional CVD-based methods to deposit β-phase Ga_2O_3 thin films still presents several challenges [4]. These include the selection of suitable precursors, precise control over thin-film thickness, and managing the potential formation of reaction byproducts during the deposition process.

Using WBG semiconductors to manufacture devices offers several significant advantages, particularly in high-performance, energy management, and extreme-condition applications [5,6]. These benefits include the following:

1. Higher operating temperature range: WBG materials such as GaN and SiC have superior thermal stability, enabling them to operate at higher temperatures than traditional Si-based semiconductors. This makes them ideal for high-temperature environments where traditional semiconductors may fail.
2. Higher switching frequencies: WBG semiconductors can operate at much higher switching speeds, which makes them particularly suitable for high-frequency power electronics applications. This includes RF amplifiers, wireless communication systems, electric vehicle (EV) chargers, and power converters, where fast switching is crucial for efficiency and performance.
3. Higher efficiency: WBG semiconductors exhibit lower conduction losses and switching losses compared with conventional silicon. In power converters, motor drive systems, and regenerative braking systems, this translates into significantly reduced energy loss, improving the overall system efficiency and reducing energy consumption.
4. Higher voltage and current handling capability: The intrinsic properties of WBG semiconductors enable them to withstand much higher voltages and currents than silicon. This makes them especially well-suited for applications requiring high voltage or high current, such as power transmission, electric vehicles, and solar energy systems.

Because WBG semiconductors provide notable improvements in system performance, energy efficiency, and reliability, they are ideal for next-generation applications in power electronics, electric vehicles, renewable energy, and beyond [7]. In the past, various methods have been explored for depositing Ga_2O_3 thin films, even those not in the β phase [8]. For example, Aleksandrova et al. used radio frequency sputtering with a commercially available target (ZnO/Ga_2O_3 = 95/5 wt.%) to deposit ZnO-doped Ga_2O_3 thin films [9], while Liu et al. employed capacitively coupled plasma-assisted magnetron sputtering to deposit amorphous Ga_2O_3 thin films [10]. However, the use of an electron beam (e-beam) system for depositing Ga_2O_3 thin films has been relatively limited. This study primarily focused on using the e-beam method to deposit Ga_2O_3 thin films. The key advantages of e-beam evaporation include high deposition rates, high-quality thin films, and high-purity thin films, as the material is typically evaporated in a vacuum. It also allows for precise control over the thickness of the deposited thin films, as the deposition rate can be carefully regulated by adjusting the e-beam power. E-beam evaporation is versatile and can be used to deposit a wide range of materials, including metals, alloys, ceramics, and some organic compounds. In the case of the NiO-doped Ga_2O_3 thin films we use, NiO and Ga_2O_3 are co-evaporated during the deposition process, allowing for the possibility of naturally forming compounds. In order to use Ga_2O_3 thin films for semiconductor device fabrication, many studies have focused on doping Ga_2O_3 with various impurities to create n-type semiconductor thin films [11,12].

Meanwhile, finding effective acceptor dopants to achieve p-type Ga_2O_3 thin films remains a challenging and active area of research [13,14]. To achieve wider bandgap p-type semiconductive Ga_2O_3 thin films, this research used NiO as the dopant material and introduced an alternative approach for preparing the target material that differs from conventional methods. This is also one of the key reasons for the ongoing interest in depositing β-phase Ga_2O_3 thin films, which further enhance these properties. The main innovation of this research lies in developing a novel fabrication technique that enables Ga_2O_3 thin films to exhibit higher WBG semiconductor properties even without the β-phase structure. Additionally, this approach allows for the straightforward induction of p-type semiconductor characteristics in Ga_2O_3 thin films. This study presents three key innovations. The first innovation lies in the development of a simple method to fabricate p-type NiO-doped Ga_2O_3 thin films. The second innovation involves a pre-sintering process, in which Ga_2O_3 powder is initially heated to the β phase before the thin films are deposited. The annealing process can enhance the properties of Ga_2O_3 thin films, such as by increasing the bandgap [15,16]. Subsequently, an annealing step is used to enhance the crystalline and electrical properties of the deposited Ga_2O_3 thin films, allowing them to exhibit the characteristics of β-phase Ga_2O_3 and thereby increasing the bandgap of the NiO-doped Ga_2O_3 thin films. The third innovation focuses on utilizing the annealing process to improve the conductivity of the NiO-doped Ga_2O_3 thin films. This is further investigated by conducting a comprehensive X-ray photoelectron spectroscopy (XPS) analysis of the e-beam-deposited NiO-doped Ga_2O_3 thin films, examining the $Ga_{2p3/2}$, Ga_{3d}, $Ni_{2p3/2}$, and O_{1s} peaks before and after annealing. The changes observed in these peaks are analyzed to explore the underlying mechanisms that contribute to the enhancement in conductivity.

2. Materials and Methods

The n-type silicon substrate (100) was used as the base material. Grease, contaminants, and organic substances adhering to the substrate surface were removed with acetone. Residual acetone was eliminated using isopropanol, followed by rinsing with deionized water to remove any remaining solvents. After completing these steps, it was essential to quickly dry the polished substrate surface with a nitrogen gun to prevent watermarks that could impact subsequent processing yield. The Ga_2O_3 powder was pre-sintered at 950 °C for 1 h. Next, the appropriate quantities of Ga_2O_3 and 12 at% NiO powders were separately added to a ball mill grinding bowl, thoroughly mixed, and then dried. The main reason for selecting 12 at% NiO as the dopant for Ga_2O_3 thin films was based on our previous electrical analyses of Ga_2O_3 thin films doped with 4 at% and 8 at% NiO. We found that the as-deposited NiO-doped Ga_2O_3 thin films were non-conductive, while the as-deposited 12 at% NiO-doped Ga_2O_3 thin films exhibited conductivity. Since our primary focus was to study the effects of annealing on NiO-doped Ga_2O_3 thin films, we chose 12 at% NiO as the doping concentration for these thin films.

We compressed the dried 12 at% NiO-doped Ga_2O_3 powder into discs with a diameter of 12 mm and a thickness of approximately 1.5 mm, followed by hardening the target at 500 °C. After the target was hardened, it was subjected to a simple crushing process to form particles with diameters ranging from 0.5 mm to 2 mm. These particles were then placed in a Mo crucible for thin-film deposition. After the sintered targets were placed in the crucible of the e-beam, vacuum evacuation commenced. The pressure in the deposition chamber was reduced to 7.8×10^{-6} torr. Subsequently, NiO-doped Ga_2O_3 thin films were deposited at room temperature. During the deposition process, only the deposition rate was controlled, while the deposition pressure was not. The deposition lasted approximately 15 min, resulting in NiO-doped Ga_2O_3 thin films with a thickness of about 200 nm. For the annealing process, the samples were transferred to a precision high-temperature sintering furnace, where they were heated in air at a rate of 7 °C/min until reaching 500 °C. They were held at this temperature for 30 min before cooling naturally to room temperature.

The primary reason for using indium as the electrode is its ability to readily melt at low temperatures, facilitating the formation of an ohmic contact with the semiconductor thin film. The procedure involves placing the sample on a hot plate, cutting an indium foil to dimensions of 0.2 cm × 0.2 cm, and positioning it on top of the thin film. The indium is then heated to 190 °C, causing it to melt and establish an ohmic contact with the thin film, as illustrated in Figure 1. After deposition, the NiO-doped Ga_2O_3 thin films were divided into two parts: one part underwent direct characterization analyses, while the other was annealed in air at 500 °C before the characterization analyses. The analyses included the following:

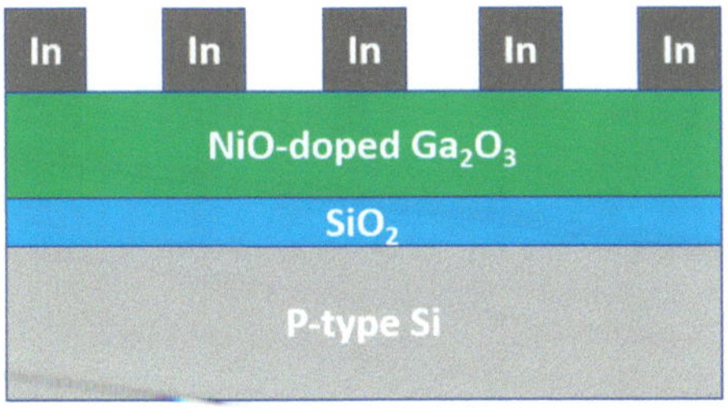

Figure 1. Schematic diagram for the NiO-doped Ga_2O_3 thin films.

(1) FESEM (field emission scanning electron microscopy) was used to observe the surface morphology and measure the thickness of the thin films.

(2) EDS (energy dispersive spectroscopy) was integrated with FESEM to analyze the distribution and concentrations of elements, confirming the successful doping of the NiO in the Ga_2O_3 thin films.

(3) XRD (X-ray diffraction) was employed to examine the crystal phases of the NiO-doped Ga_2O_3 thin films.

(4) UV-Vis spectroscopy was utilized to calculate the bandgap by analyzing the transmission spectra. For this measurement, the NiO-doped Ga_2O_3 thin films were deposited on translucent sapphire substrates.

(5) XPS (X-ray photoelectron spectroscopy) was used to analyze the photoelectrons emitted from atomic inner layers to determine the surface element composition, oxidation states, and chemical bonding of the NiO-doped Ga_2O_3 thin films.

(6) Following these characterizations, electrical property analyses were conducted using the Agilent B1500A to measure the I-V characteristics of the thin films with deposited electrodes. The measurement range was set from −2V to 2V to obtain the I-V characteristic curve. Additionally, Hall measurements were performed to assess the carrier concentration, mobility, and resistivity of the NiO-doped Ga_2O_3 thin films. The schematic diagram to obtain the various NiO-doped Ga_2O_3 thin films and the different analyses for the various NiO-doped Ga_2O_3 thin films is shown in Figure 2.

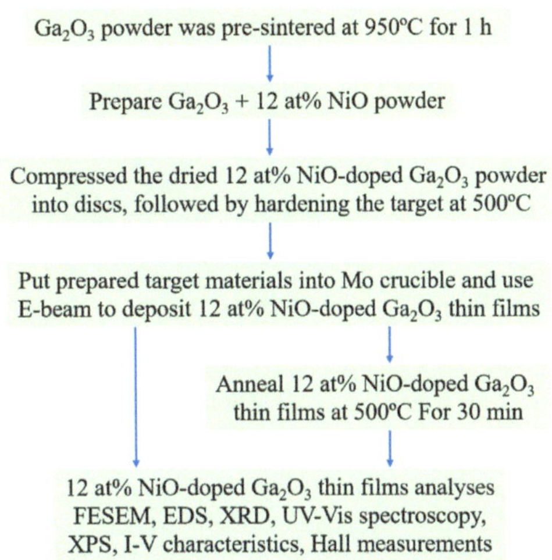

Figure 2. Schematic diagram to obtain the various NiO-doped Ga_2O_3 thin films and the different analyses for the various NiO-doped Ga_2O_3 thin films.

3. Results and Discussion

The NiO-doped Ga_2O_3 thin films were processed using different methods, leading to distinct characteristics in their surface morphologies. Figure 3 illustrates these variations, showcasing the surface morphologies of the thin films produced under different conditions. Specifically, Figure 3a displays the as-deposited NiO-doped Ga_2O_3 thin films that did not undergo annealing, while Figure 3b presents the thin films that were annealed at 500 °C. In the SEM images of the non-annealed samples, it is clear that the NiO-doped Ga_2O_3 thin films were successfully deposited via e-beam evaporation. These thin films exhibit a smooth surface, with densely packed particles, but clearly show cracks and voids on the surface. The use of e-beam deposition for thin films can lead to uneven film structures, with certain areas forming cracks or pores. These defects are typically caused by factors such as temperature variations during the deposition process, unstable deposition rates, or uneven gas distribution. During the film deposition on the substrate, internal stresses may accumulate. If these stresses exceed the yield strength of the film material, they can lead to the formation of cracks or pores. The stresses usually arise from thermal expansion mismatch, differences in the thermal expansion coefficients between the substrate and the film, or excessive film thickness. This is why the deposited thin film, specifically on the surface, exhibits cracks and voids.

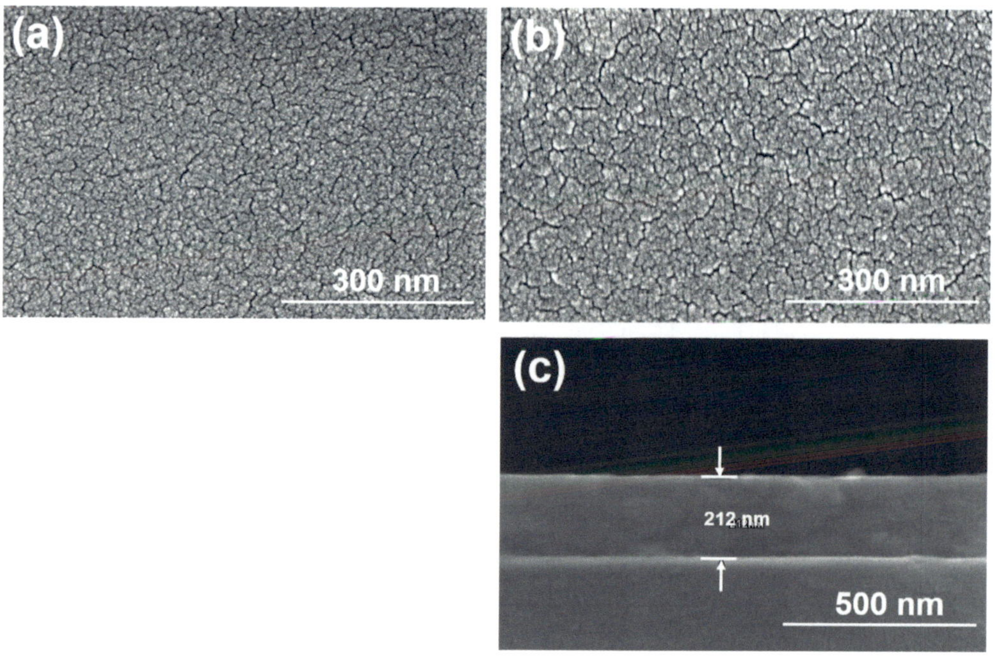

Figure 3. Surface morphologies of (**a**) un-annealed and (**b**) 500 °C-annealed NiO-doped Ga_2O_3 thin films and (**c**) the cross-sectional image of the deposited NiO-doped Ga_2O_3 thin films.

However, as seen in Figure 3a, there is an absence of visible microcrystals in the deposited films. In contrast, the annealing process at 500 °C did not induce cracks or pores due to thermal contraction, which contributed to an increase in film density. The regions of the thin films that remain smooth still exhibit excellent density, indicating a robust structure. Additionally, the annealing treatment facilitated the formation of the microcrystalline structure, which is clearly visible in Figure 3b for the 500 °C-annealed NiO-doped Ga_2O_3 thin films. This transformation highlights the significant impact of thermal treatment on the structural properties of the NiO-doped Ga_2O_3 thin films. The most significant findings presented in Figure 3a,b indicate that neither the as-deposited NiO-doped Ga_2O_3 thin films nor the NiO-doped Ga_2O_3 thin films that were annealed at 500 °C exhibited any abnormal deposits or signs of film delamination on their surfaces. This suggests that the deposition process and subsequent thermal treatment resulted in stable and uniform films that are free from defects that could compromise their structural integrity. Figure 3c displays cross-sectional images used to measure the thickness of the as-deposited NiO-doped Ga_2O_3 thin films. The measured thickness of these films was 215 nm, providing insight into their structural dimensions and confirming the successful deposition process.

Figure 4a shows the EDS analysis of the 500 °C-annealed NiO-doped Ga_2O_3 thin films. Similar results were obtained for the un-annealed NiO-doped Ga_2O_3 thin films, and therefore, the analysis is not presented here. The EDS spectra confirm the successful detection of Ga, Ni, and O elements in the 500 °C-annealed NiO-doped Ga_2O_3 thin films, as shown in Figure 4b. The average ratios of the Ga, Ni, and O elements for the deposited NiO-doped Ga_2O_3 thin films and the 500 °C-annealed NiO-doped Ga_2O_3 thin film were 42.54:5.98:51.48 and 41.15:5.74:53.11, respectively. The Ga-to-O ratio is not 2:3, primarily due to the presence of NiO, which causes the Ga-to-O ratio to be lower than 2:3. This result demonstrates the effective incorporation of NiO into the Ga_2O_3 matrix, affirming the viability of the doping process and its successful implementation in modifying the thin film composition.

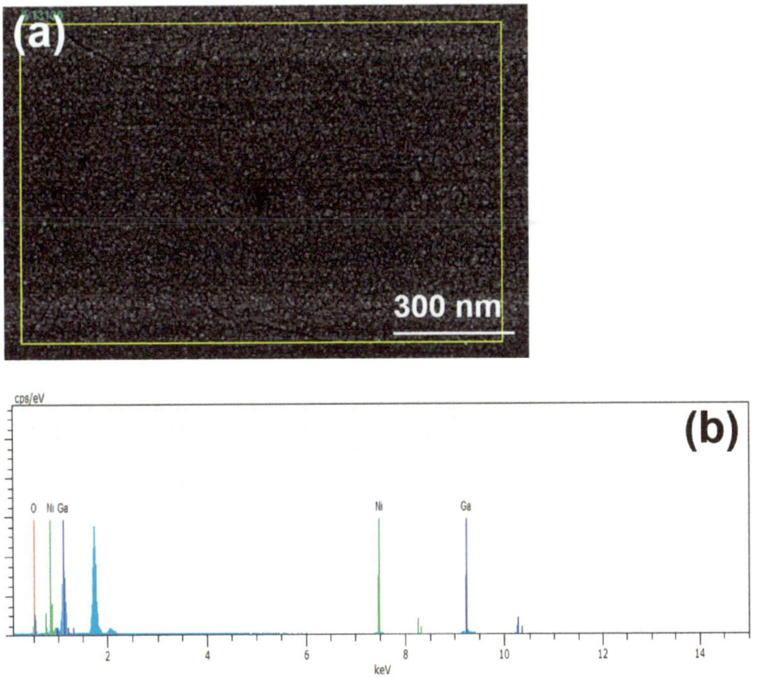

Figure 4. EDS analysis of the 500 °C-annealed NiO-doped Ga_2O_3 thin films: (**a**) analysis area and (**b**) analysis result.

The XRD spectrum reveals the crystalline properties of the thin films subjected to different annealing conditions, with the peaks indicating structural and phase changes in the thin films. The X-ray diffraction (XRD) analyses revealed that both the un-annealed and the 500 °C-annealed NiO-doped Ga_2O_3 thin films exhibited an amorphous phase. In contrast, the Ga_2O_3 powder pre-sintered at 950 °C showed distinct diffraction peaks corresponding to the β-Ga_2O_3 phase (not shown here). The main reason for the persistence of the amorphous phase in the 500 °C-annealed NiO-doped Ga_2O_3 thin films is that the powder was pre-heated to 950 °C before deposition, causing it to transform into the β-Ga_2O_3 phase. When the powder was subsequently evaporated by the electron beam deposition process, it re-condensed into fine particles and was deposited in an amorphous state. Even after annealing at 500 °C, the crystal structure of the pre-sintered Ga_2O_3 powder remained stable, with its primary phase still being Ga_2O_3. As a result, no new crystalline phases were formed during the annealing process, and the material remained amorphous. The transmittance spectra illustrate the material's interaction with light, offering insights into its bandgap properties. Variations in the annealing conditions can affect the absorption edges and shifts, reflecting changes in the bandgap.

Measurements of the transmittance spectrum show that both the un-annealed and 500 °C-annealed NiO-doped Ga_2O_3 thin films exhibited distinct absorption edges. Notably, after post-annealing, a blue shift in the absorption edge was observed, indicating an increase in the bandgap for the 500 °C-annealed NiO-doped Ga_2O_3 thin films. Previous studies found that while high-temperature annealing in a sintering furnace did not significantly alter the crystalline properties of Ga_2O_3 thin films, it did have a substantial impact on the Eg values (energy bandgap) of the deposited Ga_2O_3 thin films [17]. In the past, various methods had been employed to calculate the Eg values of certain materials [15]. This study primarily utilized transmittance spectra data, applying the Tauc plot to determine the Eg values. This approach aligns with the observations presented in Figure 5, where variations in the bandgap are clearly evident. The determined bandgaps for the NiO-doped Ga_2O_3 thin films are illustrated in Figure 5a,b, corresponding to the un-annealed and 500 °C-annealed samples, respectively. It is well established that the annealing process can enhance the crystallinity of materials, leading to a reduction in defect states within the bandgap and ultimately affecting their Eg values. The calculated Eg values for the un-annealed and 500 °C-annealed NiO-doped Ga_2O_3 thin films are 3.878 eV and 4.402 eV, respectively. This significant increase in the bandgap following annealing clearly demonstrates the positive impact of the annealing process on the crystallinity of the NiO-doped Ga_2O_3 thin films, highlighting the material's improved electronic properties.

The primary reason for this result is that the powder underwent a pre-sintering process at 950 °C, which already converted the powder into the β-Ga_2O_3 phase. After deposition, the resulting NiO-doped Ga_2O_3 thin films are amorphous. Although annealing at 500 °C does not fully crystallize the thin films into the β-Ga_2O_3 phase, and the thin films remain predominantly amorphous, it does promote the formation of microcrystals within the Ga_2O_3 thin films. We believe that the 500 °C annealing process activates the characteristics of the β-Ga_2O_3 phase. Even though the 500 °C-annealed thin films do not completely crystallize into β-Ga_2O_3, the heat treatment seems to stimulate local regions of the NiO-doped Ga_2O_3 thin films that exhibit β-phase-like properties. This partial crystallization likely contributes to the observed increase in the Eg value of the Ga_2O_3 thin films. To further validate the effects of annealing on NiO-doped Ga_2O_3 thin films, XRD analyses and the calculation of the Eg value for undoped and as-deposited Ga_2O_3 thin films were also conducted. The XRD analyses revealed that the undoped and as-deposited Ga_2O_3 thin films exhibited an amorphous phase. Figure 5c illustrates the Eg value calculation for the undoped and as-deposited Ga_2O_3 thin films, which was determined to be 3.786 eV. This value closely aligns with the Eg value of the as-deposited NiO-doped Ga_2O_3 thin films. Consequently, this observation further substantiates that the increase in the Eg value of the 500 °C-annealed NiO-doped Ga_2O_3 thin films is primarily attributed to the effects of the 500 °C annealing process.

Figure 6 presents the SIMS analysis results for the elements Ga, Ni, Si, and O in Figure 6a for the un-annealed NiO-doped Ga_2O_3 thin films and in Figure 6b for the 500 °C-annealed NiO-doped Ga_2O_3 thin films. Figure 6 shows that there is no significant difference in the elemental changes between the un-annealed NiO-doped Ga_2O_3 thin films and the 500 °C-annealed NiO-doped Ga_2O_3 thin films. For example, in the case of the 500 °C-annealed NiO-doped Ga_2O_3 thin films, the analysis results revealed that the concentration of Ni shows a gradual decrease and that of O is almost unchanged from the surface to the deeper layers of the thin film. Notably, at a depth of approximately 225 nm, both Ga and Ni concentrations decline sharply. In contrast, the concentration of Si begins to increase slightly around 180 nm, reaching its peak at approximately 215 nm. This trend indicates a distinct compositional variation within the layers of the NiO-doped Ga_2O_3 thin films, highlighting the interactions between these elements as the depth increases. Conversely, the concentration of Ga exhibits minimal variation from the surface down to a depth of 225 nm. In terms of O, its concentration is unchanged from the surface; however, a more pronounced decrease begins around the 215 nm mark. The observed increase in Si concentration can be attributed to the analysis depth reaching the underlying Si substrate, suggesting an interaction with the thin film. The results of these analyses confirm the accuracy of the thickness observations from the SEM image of the NiO-doped Ga_2O_3 thin films.

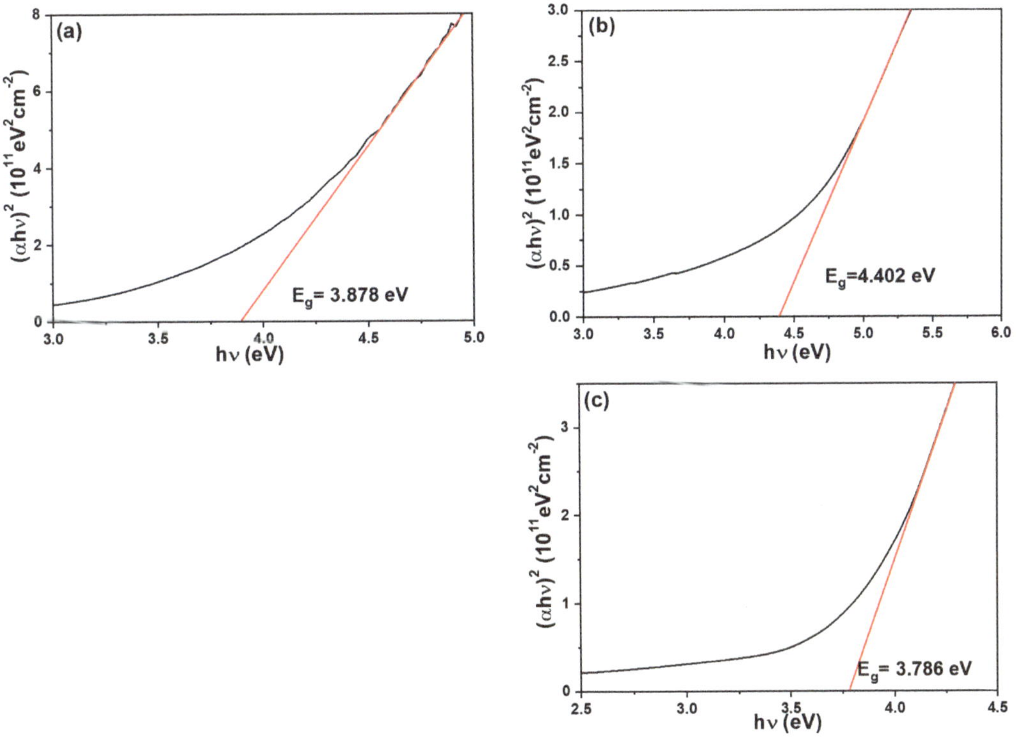

Figure 5. Tauc plots of (**a**) un-annealed, (**b**) 500 °C-annealed NiO-doped Ga_2O_3 thin films, and (**c**) un-annealed and undoped Ga_2O_3 thin films.

Additionally, these results closely match the measured thickness of the deposited NiO-doped Ga_2O_3 thin films, approximately 215 nm. This consistency further reinforces the reliability of both the imaging technique and the thickness measurements. The melting temperatures of Ga_2O_3 and NiO powders are 1725 °C and 1955 °C, respectively. Furthermore, the gradual decrease in Ni concentration and the uniform distribution of Ga concentration indirectly support the earlier explanation that NiO evaporates more slowly during deposition, while the Ga_2O_3 components tend to evaporate first. This behavior underscores the differing thermal stabilities of the two materials throughout the deposition process. As a result, the concentration of Ni diminishes from the bottom of the NiO-doped Ga_2O_3 thin films. Furthermore, the nearly constant oxygen (O) content observed from the surface inward suggests that there is no significant decomposition of NiO into its constituent elements, Ni and oxygen, during the deposition of the NiO-doped Ga_2O_3 thin films. This stability indicates a well-maintained chemical integrity of the thin films, which is crucial for their intended applications.

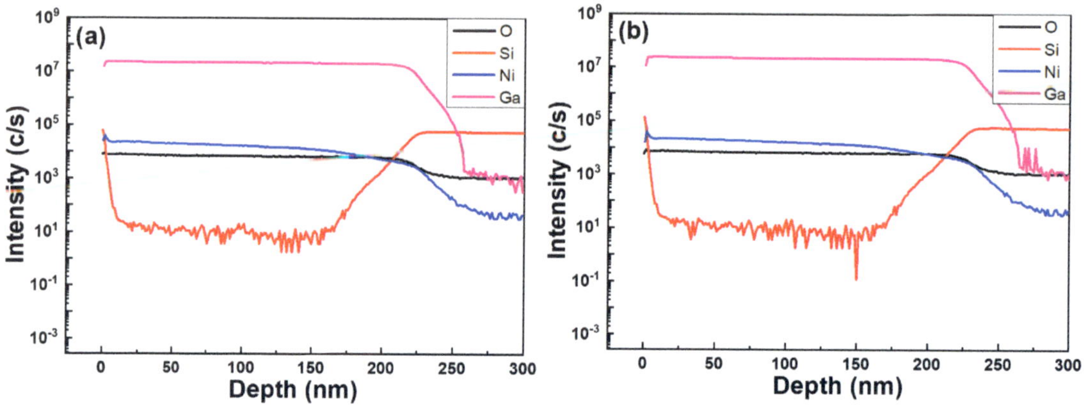

Figure 6. SIMS analysis results of Ga, Ni, Si, and O elements of (**a**) un-annealed NiO-doped Ga_2O_3 thin films and (**b**) 500 °C-annealed NiO-doped Ga_2O_3 thin films.

This study aimed to enhance the conductivity of the deposited Ga_2O_3 thin films, which inherently lack electrical conductivity, by doping them with NiO. According to the I-V measurement curves, we found that pure Ga_2O_3 thin films showed no conductivity, making it nearly impossible to observe significant changes in the I-V characteristics, as illustrated in Figure 7. Therefore, we have chosen not to present the measurement results of the pure Ga_2O_3 thin films. In contrast, we discovered that both types of the NiO-doped Ga_2O_3 thin films prepared through different methods exhibited notable conductivity. We proceeded to analyze and compare their electrical performances. The study evaluated an un-annealed NiO-doped Ga_2O_3 thin film alongside a NiO-doped Ga_2O_3 thin film that was annealed at 500 °C, using a B1500A electrical analysis system. The experimental results, displayed in Figure 7, revealed that within the measurement range of −2V to +2V, both the un-annealed and the 500 °C-annealed NiO-doped Ga_2O_3 thin films, each approximately 212 nm thick, demonstrated a measurable level of conductivity. This finding indicates that doping Ga_2O_3 with NiO effectively enhances its electrical properties, opening new avenues for its application in electronic devices.

In our experiments, we investigated the I-V characteristics of the NiO-doped Ga_2O_3 thin films under applied bias voltages of −2V and +2V. When measuring the un-annealed NiO-doped Ga_2O_3 thin film, we observed that the maximum current density reached approximately 10^{-4} A/cm^2. In contrast, using the NiO-doped Ga_2O_3 thin film that had been annealed at 500 °C under the same bias conditions, we recorded a significantly higher maximum current density of around 3.2×10^0 A/cm^2. This stark difference highlights the impact of thermal annealing on the electrical properties of the NiO-doped Ga_2O_3 thin films. Pure Ga_2O_3 thin films typically exhibit n-type semiconductor behaviors, which arise from the defects or impurities within the material itself. These defects or impurities introduce additional electrons within the bandgap, resulting in n-type characteristics. Several factors may contribute to the emergence of p-type semiconductor behavior in NiO-doped Ga_2O_3 thin films. NiO is a well-known p-type semiconductor material, and its p-type characteristics can be maintained or even enhanced when Ga_2O_3 thin films are added, through several mechanisms.

First, Ni^{2+} has an ionic radius of approximately 0.069 nm, while Ga^{3+} has a slightly smaller ionic radius of about 0.062 nm. Nickel, being a common transition metal, has an ionic radius that is quite similar to that of gallium, suggesting its potential to substitute for Ga atoms within the Ga_2O_3 lattice. NiO itself is a p-type semiconductor, and the doped nickel ions (Ni^{2+}) contribute to the generation of holes. When NiO is incorporated into Ga_2O_3, these holes can participate in electrical conduction, thereby imparting p-type characteristics to the overall film. Moreover, the introduction of NiO alters the band structure of Ga_2O_3, potentially leading to the formation of new energy bands or gaps. This modification can promote the generation and movement of holes, further enhancing the p-type conductivity of the material. Therefore, the combination of NiO with Ga_2O_3 not only leverages the similar ionic radii of Ni^{2+} and Ga^{3+} for effective lattice substitution but also modifies the electronic properties of the host material, facilitating improved p-type conduction. In this study, we demonstrate the transformation of completely non-conductive Ga_2O_3 thin films into conductive NiO-doped Ga_2O_3 thin films, highlighting the feasibility and potential impact of our research. This transition not only emphasizes the effectiveness of doping but also opens up new possibilities for applications in electronic devices.

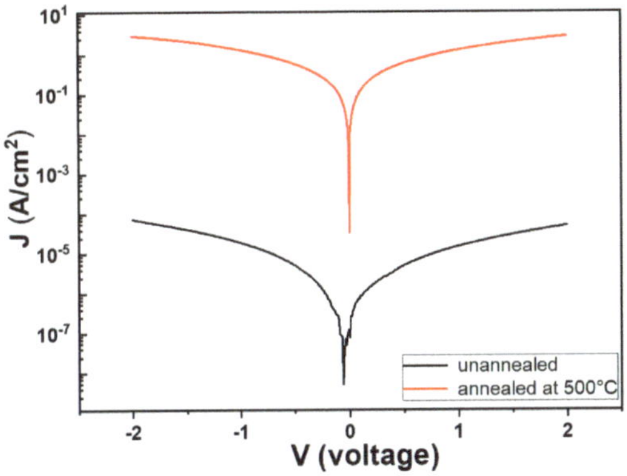

Figure 7. Current–voltage properties of the un-annealed and 500 °C-annealed NiO-doped Ga_2O_3 thin films.

The assessment of the current–voltage characteristics demonstrated that both un-annealed and 500 °C-annealed NiO-doped Ga_2O_3 thin films exhibit remarkably favorable electrical properties. To gather further data, Hall effect measurements were performed on both sets of thin films. The results revealed that the carrier concentration in the un-annealed NiO-doped Ga_2O_3 thin films was $1.42 \times 10^{17}/cm^3$, with a mobility of 3.20×10^1 cm^2/Vs and a resistivity of 2.42×10^0 $\Omega \cdot cm$. In contrast, the 500 °C-annealed NiO-doped Ga_2O_3 thin films showed a significantly higher carrier concentration of $5.35 \times 10^{18}/cm^3$, a mobility of 2.74×10^2 cm^2/Vs, and a much lower resistivity of 5.61×10^{-3} $\Omega \cdot cm$. These findings underscore a crucial point: the 500 °C-annealed NiO-doped Ga_2O_3 thin films demonstrate superior conductivity compared with their un-annealed counterparts, as evidenced by the larger current values presented in Figure 7. This improvement in electrical properties can be attributed to the annealing process, which enhances the impurity concentration and mobility within the NiO-doped Ga_2O_3 thin films. The annealing step facilitates the deeper incorporation of NiO into the Ga_2O_3 matrix, thereby amplifying the previously mentioned effects and ultimately leading to increased conductivity. The electrical conductivity of the thin films was calculated based on the measured values, confirming that NiO-doped Ga_2O_3 thin films annealed at 500 °C exhibited excellent conductivity. The Hall parameters, including carrier concentration and mobility, are influenced by several factors, such as the thickness, doping concentration, material purity, and fabrication processes of the deposited thin films. These factors significantly impact the electrical properties of the NiO-doped Ga_2O_3 thin films, underscoring the complexity of optimizing their conductivity for practical applications.

XPS is a powerful quantitative spectroscopic method used to analyze the elemental composition of materials at the parts-per-thousand level as well as to assess the electronic configurations and the chemical bonding states of the elements present in deposited thin films. This method is invaluable for understanding not only which elements are contained in deposited films but also the nature of their interactions with other elements. For NiO-doped Ga_2O_3 thin films, the decrease in conductivity is closely related to changes in different XPS peaks. In our study, we employed XPS to investigate the chemical bonding states of oxygen (O), gallium (Ga), and nickel (Ni) in the un-annealed NiO-doped Ga_2O_3 thin films and the 500 °C-annealed NiO-doped Ga_2O_3 thin films. By analyzing the measured data, we aimed to identify the factors contributing to variations in the electrical properties of two different NiO-doped Ga_2O_3 thin films.

The typical $Ga_{2p3/2}$ peaks of the un-annealed NiO-doped Ga_2O_3 thin films and 500 °C-annealed NiO-doped Ga_2O_3 thin films, centered at 1118.03 and 1118.63 eV, respectively, are shown in Figure 8a,b. For the un-annealed NiO-doped Ga_2O_3 thin films (500 °C-annealed NiO-doped Ga_2O_3 thin films), this peak consisted of three components: the Ga_I peak at 1117.96 (1117.40) eV, attributed to metallic Ga-Ga (or Ga-Ni); the Ga_{II} peak at 1118.71 (1118.15) eV, associated with the Ga_2O_3 structure; and the Ga_{III} peak at 1119.40 (1118.86) eV, corresponding to the Ga-O bond or high-oxidation-state Ga species formed on the surface or interface, such as $Ga(OH)_3$ or hydrated oxides [18]. In p-type semiconductors, conductivity primarily depends on the holes' concentration. Therefore, any alteration in the chemical environment that weakens the supply or mobility of holes will result in a decrease in conductivity. The enhancements of the $Ga(OH)_3$ or Ga-OH peaks (representing surface-adsorbed water or oxides) may contribute to this conductivity drop. This is because surface species such as $Ga(OH)_3$ can influence the carrier concentration in the surface layer, reducing the effective hole concentration and, consequently, lowering the material's conductivity. As shown in Figure 8, when NiO-doped Ga_2O_3 thin films are annealed at 500 °C, the area of the Ga_{III} peak decreases from 10.224% to 7.864%, which demonstrates that the 500 °C-annealed NiO-doped Ga_2O_3 thin films can exhibit improved conductivity. The typical $Ga_{2p1/2}$ peaks of the un-annealed NiO-doped Ga_2O_3 thin films and the 500 °C-annealed NiO-doped Ga_2O_3 thin films are compared in Figure 9, and their binding energies are centered at 1147.70 and 1145.36 eV, respectively, as shown in Figure 9a,b. By comparing the analysis results of Figures 9 and 10, we observe that the XPS trend of the $Ga_{2p1/2}$ peaks is similar to that of the $Ga_{2p3/2}$ peaks. As shown in Figure 9, when NiO-doped Ga_2O_3 thin films are annealed at 500 °C, the area of the Ga_{III} peak decreases from 10.568% to 8.103%, which also demonstrates that the 500 °C-annealed NiO-doped Ga_2O_3 thin films can exhibit improved conductivity.

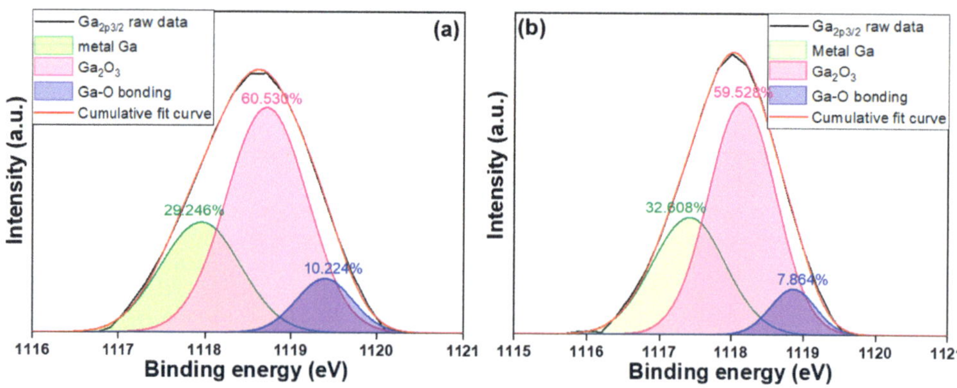

Figure 8. XPS spectra of the $Ga_{2p3/2}$ peaks and the Gaussian-resolved components (**a**) un-annealed NiO-doped Ga_2O_3 thin films and (**b**) 500 °C-annealed NiO-doped Ga_2O_3 thin films.

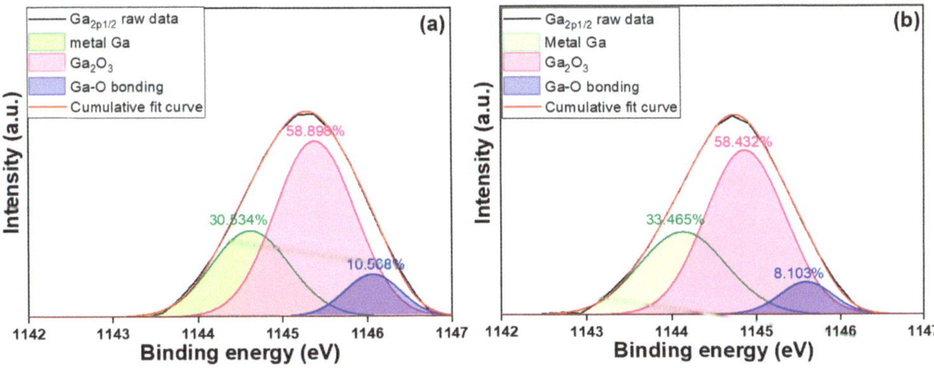

Figure 9. XPS spectra of the $Ga_{2p1/2}$ peaks and the Gaussian-resolved components (**a**) un-annealed NiO-doped Ga_2O_3 thin films and (**b**) 500 °C-annealed NiO-doped Ga_2O_3 thin films.

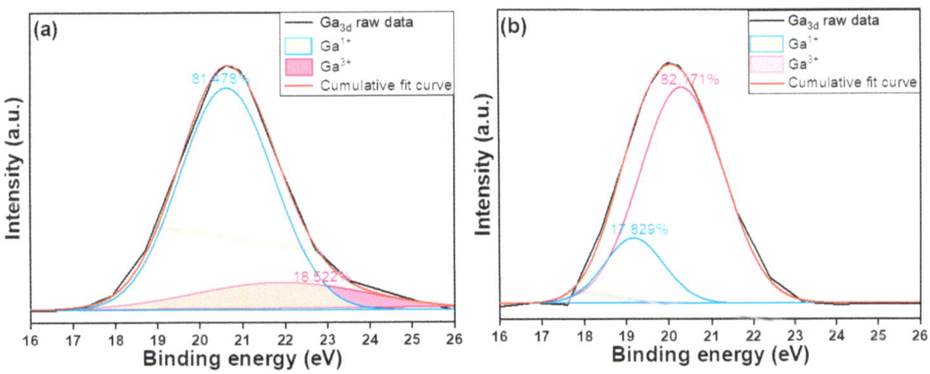

Figure 10. XPS spectra of the Ga_{3d} peaks and the Gaussian-resolved components in (**a**) un-annealed NiO-doped Ga_2O_3 thin films and (**b**) 500 °C-annealed NiO-doped Ga_2O_3 thin films.

The Ga_{3d} peak can be divided into two components: the Ga^{1+} peak and the GaO bonding peak. The Ga^{1+} peak is typically associated with oxygen vacancies or other defects, indicating that Ga atoms have lost some of their positive charge, possibly forming structures like Ga_2O [19]. This peak suggests that a portion of the gallium is in a reduced state, which can impact the overall properties of the material. Conversely, the GaO bonding peak represents the typical Ga-O bond, where gallium predominantly exists in the form of Ga^{3+}. This reflects the normal chemical bonding between gallium and oxygen atoms, indicating a stable and well-formed network within the material. Figure 10a shows that the Ga_{3d} peak of an un-annealed NiO-doped Ga_2O_3 thin film was centered at 20.783 eV, and the Ga_{3d} peak of a 500 °C-annealed NiO-doped Ga_2O_3 thin film was centered at 20.000 eV, respectively. In NiO-doped p-type Ga_2O_3 thin films, an increase in the Ga^{1+} peak can lead to a decrease in the thin film's conductivity. NiO primarily functions as an acceptor in this context by substituting Ni^{2+} for Ga^{3+}, which creates holes that contribute to p-type conductivity. Ideally, the chemical state of Ni^{2+}-O^{2-} should be maintained to ensure effective generation of holes. However, the presence of the Ga^{1+} peak introduces negative effects that reduce conductivity. Specifically, the existence of Ga^{1+} suggests that some gallium is in a reduced state, which tends to form stronger bonds with surrounding O^{2-} ions.

This alteration can disrupt the local charge balance, thereby interfering with the effectiveness of Ni^{2+} as an acceptor. As a result, the effective concentration of available acceptors can generate a decrease in the holes. Therefore, an increase in the presence of Ga^{1+}

states negatively impacts the performance of the NiO-doped p-type Ga_2O_3 semiconductor thin films. This phenomenon highlights the complex interplay between defect states and conductivity in semiconductor materials, emphasizing the need for careful control of doping and defect concentrations to optimize the electronic properties of the NiO-doped p-type Ga_2O_3 thin films. Figure 10 shows that after the annealing step at 500 °C for the NiO-doped p-type Ga_2O_3 thin films, the area of the Ga^{1+} peak significantly decreases, as shown in Figure 10a, while the area of the Ga-O bond peak substantially increases, as shown in Figure 10b. This observation indicates an enhancement in conductivity. The reduction in the Ga^{1+} peak suggests a decrease in the presence of reduced gallium states, which can interfere with the generation of holes. Meanwhile, the increase in the Ga-O bond peak reflects improved Ga-O bonding and a more stable structure, contributing to enhanced electronic transport properties in the material. Overall, these changes are favorable for improving the conductivity of the NiO-doped Ga_2O_3 semiconductor thin films.

In XPS analysis, the splitting of the $Ni_{2p3/2}$ peak typically indicates the presence of different chemical environments or oxidation states. Specifically, the two components of the $Ni_{2p3/2}$ peak may correspond to nickel atoms in different chemical states, which could arise from varying oxidation states of nickel or its bonding with other elements. The Ni^{2+} oxidation state is commonly observed in NiO, where nickel primarily exists as Ni^{2+} ions. Thus, this peak is often associated with the electronic structure of Ni^{2+} ions in NiO. On the other hand, the Ni^{3+} oxidation state may appear in certain situations, particularly when nickel is doped into the Ga_2O_3 lattice. In these cases, nickel may exist as Ni^{3+} ions, which have a distinct electronic structure and typically manifest as a separate split peak in the XPS spectrum. In Ga_2O_3 thin films doped with NiO, the presence of Ni^{2+} ions usually has a stronger promoting effect on the p-type properties. Ni^{2+} ions act as an acceptor dopant, introducing holes into Ga_2O_3 and thereby enhancing its p-type conductivity. In contrast, the presence of Ni^{3+} ions could negatively impact p-type conductivity. The relatively stable electronic structure of Ni^{3+} ions might reduce the generation or mobility of holes, thus impairing the conductive properties of the material.

Figure 11a,b show the XPS analysis results of the $Ni_{2p3/2}$ and $Ni_{2p1/2}$ peaks for un-annealed and 500 °C-annealed NiO-doped Ga_2O_3 thin films, respectively. By comparing the data from Figure 11a,b, we can observe significant changes following the 500 °C annealing process. Specifically, in the $Ni_{2p3/2}$ peak, the area of the Ni^{2+} (Ni^{3+}) peaks decreased (increased) notably after annealing. However, the $Ni_{2p1/2}$ peak showed a clear opposite trend, with the area of the Ni^{2+} (Ni^{3+}) peaks increasing (decreasing) after annealing. This change could be due to the relatively weak intensity of the $Ni_{2p1/2}$ peak in the XPS analysis, which may affect the accuracy of the analysis results. These findings also suggest that, although annealing at 500 °C does not significantly promote the crystallinity of the NiO-doped Ga_2O_3 thin films (as shown in Figure 3), it does facilitate the incorporation of NiO into the Ga_2O_3 matrix to be changed to Ni^{3+} ions. While the 500 °C annealing process may increase the area of Ni^{3+} peak and slightly reduce the p-type conductivity of the NiO-doped Ga_2O_3 thin films, taking into account other factors, such as improved doping and the overall structural effects, it still leads to enhanced conductivity compared with the un-annealed thin films.

Figure 12 illustrates the O_{1s} peak of un-annealed NiO-doped Ga_2O_3 thin films and 500 °C-annealed NiO-doped Ga_2O_3 thin films. The typical O_{1s} peak can be deconvoluted into the O_I, O_{II}, and O_{III}, three nearly Gaussian components. For un-annealed NiO-doped Ga_2O_3 thin films (Figure 12a) and 500 °C-annealed NiO-doped Ga_2O_3 thin films (Figure 12b), the centers of the O_{1s} peaks are located at 530.80 eV and 531.31 eV. The centers of these peaks of un-annealed NiO-doped Ga_2O_3 thin films are located at 530.05 ± 0.03 eV (O_I), 530.72 ± 0.02 eV (O_{II}), and 531.45 ± 0.02 eV (O_{III}). The centers of these peaks of 500 °C-annealed NiO-doped Ga_2O_3 thin films are located at 530.48 ± 0.02 eV (O_I), 531.15 ± 0.02 eV (O_{II}), and 531.88 ± 0.03 eV (O_{III}), respectively. This detailed analysis of the O_{1s} peaks provides crucial insights into the chemical environment of oxygen in the NiO-doped Ga_2O_3 thin films, helping us to elucidate the underlying mechanisms that influence their electrical characteristics. Abenuz Acuña et al. found that a lower binding energy peak at O_I is associated with the NiO phase, while the peak of the intermediate binding energy at O_{II} is attributed to NiO(OH). The peak observed at the higher binding energy of O_{III} is linked to organic contamination, such as carbon, which is commonly found in most samples and easily adsorbs when exposed to the environment [20].

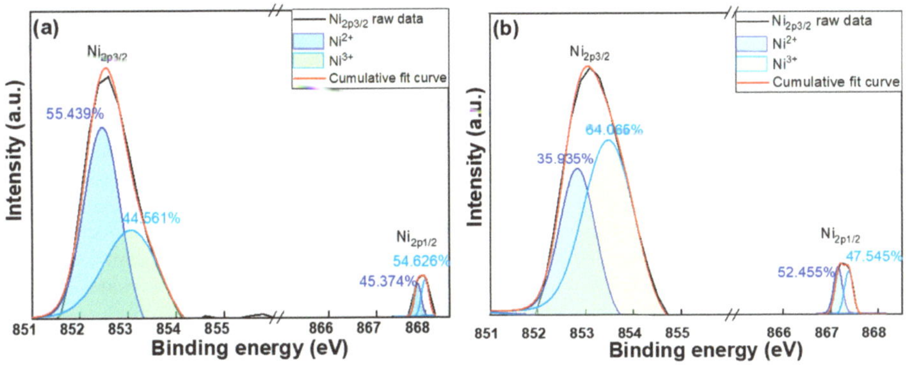

Figure 11. XPS spectra of the $Ni_{2p3/2}$ and Ni $_{2p1/2}$ peaks and the Gaussian-resolved components of (**a**) un-annealed NiO-doped Ga_2O_3 thin films and (**b**) 500 °C-annealed NiO-doped Ga_2O_3 thin films.

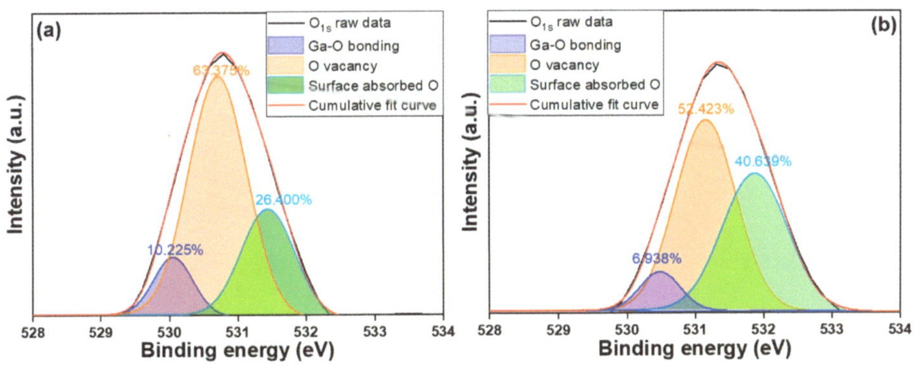

Figure 12. XPS spectra of the O_{1s} peaks and the Gaussian-resolved components in (**a**) un-annealed NiO-doped Ga_2O_3 thin films and (**b**) 500 °C-annealed NiO-doped Ga_2O_3 thin films.

Salunkhe et al. found that a lower binding energy peak at O_I is caused by the interaction of lattice oxygen in the form of Ni–octahedral bonding in NiO. The peak at the intermediate binding energy, O_{II}, is attributed to Ni^{3+}, indicating vacancies or metal deficiencies in the thin films. The peak observed at the higher binding energy, O_{III}, is linked to the presence of adsorbed O_2 on the surface layer of the deposited NiO [21]. The O_I peak is typically associated with structural oxygen (also known as lattice oxygen), which is part of the Ga_2O_3 crystal structure and significantly influences the material's fundamental properties. The O_{II} peak is related to defects or oxygen vacancies, which affect the electrical properties of the material, particularly its conductivity. The O_{III} peak is generally linked to surface oxygen or chemisorbed oxygen, which may originate from the environment or the surface characteristics of the oxide. In terms of conductivity, the O_{II} peak often has a more substantial impact [22]. The defect states of oxygen are crucial for the electronic transport properties of the NiO-doped Ga_2O_3 thin films. As the annealing temperature increased from room temperature to 500 °C, the area of the O_I peak decreased from 10.225% to 6.938%, while the area of the O_{III} peak increased from 26.400% to 40.639%. Meanwhile, the area of the O_{II} peak, which represents oxygen vacancies, decreased from 63.375% to 52.423%. Additionally, mobility improved with the increase in annealing temperature, following a trend similar to that of the O_{II} peak area.

This indicates that the enhanced conductivity of the NiO-doped Ga_2O_3 thin films after annealing at 500 °C is primarily due to the decreased area of the O_{II} peak, reflecting a decrease in oxygen vacancies. In the O_{1s} XPS analyses of the Ga_2O_3 thin films, an increase in the area of the O_{II} peak typically indicates a higher number of oxygen vacancies or defects. This situation usually leads to a change in the conductivity of the deposited NiO-doped Ga_2O_3 thin films. In n-type semiconductors, the primary charge carriers are electrons. Oxygen vacancies provide additional electrons, increasing the concentration of conductive carriers and enhancing conductivity. In p-type semiconductors, the main charge carriers are holes, but the electrons supplied by oxygen vacancies can recombine with these holes, reducing their number. Furthermore, in some cases, the presence of oxygen vacancies may introduce new energy levels that can act as traps for both electrons and holes, thereby affecting carrier mobility. Both of these factors contribute to a decrease in conductivity in p-type semiconductors. Consequently, reducing oxygen vacancies leads to an increase in conductivity. This is evidenced by Figure 12, which demonstrates that the NiO-doped Ga_2O_3 thin films annealed at 500 °C exhibit higher conductivity.

4. Conclusions

From the SEM and EDS analyses of the NiO-doped Ga_2O_3 thin films, the measured thickness of the thin films was 215 nm, with average elemental ratios of Ga, Ni, and O of 41.15:5.74:53.11. For the un-annealed NiO-doped Ga_2O_3 thin films, the carrier concentration was $1.42 \times 10^{17}/cm^3$, with a mobility of 3.20×10^1 cm^2/Vs and a resistivity of 2.42×10^0 $\Omega \cdot cm$. In contrast, the 500 °C-annealed NiO-doped Ga_2O_3 thin films exhibited a significantly higher carrier concentration of $5.35 \times 10^{18}/cm^3$, a higher mobility of 2.74×10^2 cm^2/Vs, and a much lower resistivity of 5.61×10^{-3} $\Omega \cdot cm$. For the XPS analyses, when NiO-doped Ga_2O_3 thin films were annealed at 500 °C, the area of the Ga_{III} peak in the $Ga_{2p3/2}$ peak decreased from 10.224% to 7.864%, indicating that the annealed films exhibit improved conductivity. The reduction in the Ga^{1+} peak in the Ga_{3d} peak caused a decrease in the presence of reduced gallium states, which hindered the generation of holes. Meanwhile, the increase in the Ga–O bond peak improved Ga–O bonding and a more stable structure, both of which contributed to enhanced electronic transport properties. Overall, the analysis of the Ga_{3d} peak supports the improvement in conductivity of the NiO-doped Ga_2O_3 thin films. For the NiO-doped Ga_2O_3 thin films, the acceptor dopant of the Ni^{2+} peak caused a decrease in the $Ni_{2p3/2}$ peak and an increase in the $Ni_{2p1/2}$ peak. The former indicates a reduction in p-type conductivity, while the latter indicates an enhancement of p-type conductivity. For the O_{1s} peak, the enhanced conductivity of the NiO-doped Ga_2O_3 thin films after annealing at 500 °C was primarily due to the decreased area of the O_{II} peak, reflecting a decrease in oxygen vacancies.

Author Contributions: Conceptualization, C.-F.Y., T.-H.M. and S.-H.L.; formal analysis, C.-F.Y., E.-C.T., Y.-W.W., H.-P.L., T.-H.M. and S.-H.L.; investigation, C.-F.Y., E.-C.T., T.-H.M. and S.-H.L.; data curation, C.-F.Y., E.-C.T., Y.-W.W., H.-P.L., T.-H.M. and S.-H.L.; writing—original draft preparation, C.-F.Y., T.-H.M. and S.-H.L.; writing—review and editing, C.-F.Y., T.-H.M. and S.-H.L.; visualization, C.-F.Y., E.-C.T., Y.-W.W., H.-P.L., T.-H.M. and S.-H.L. All authors have read and agreed to the published version of the manuscript.

Funding: This research is supported by projects under Nos. NSTC 113-2622-E-390-001 and NSTC 113-2221-E-390-011.

Institutional Review Board Statement: Not applicable.

Informed Consent Statement: Not applicable.

Data Availability Statement: Data are contained within the article.

Acknowledgments: This research is supported by Summit-Tech Resource Corp.

Conflicts of Interest: The authors declare no conflicts of interest.

References

1. Ma, X.; Zhang, Y.; Dong, L.; Jia, R. First-principles calculations of electronic and optical properties of aluminum-doped β-Ga_2O_3 with intrinsic defects. *Results Phys.* **2017**, *7*, 1582–1589. [CrossRef]
2. Rex, J.P.; Kwong, Y.F.; San, L.H. The influence of deposition temperature on the structural, morphological and optical properties of micro-size structures of beta-Ga_2O_3. *Results Phys.* **2019**, *14*, 102475. [CrossRef]
3. Wang, Y.; Li, J.; Zhang, T.; Wu, W.; Li, W.; Yao, Y.; Wang, Z.; Feng, Q.; Zhang, Y.; Zhang, J.; et al. Enhancing the quality of homoepitaxial (2201) β-Ga_2O_3 thin film by MOCVD with in situ pulsed indium. *Appl. Phys. Lett.* **2024**, *124*, 072105. [CrossRef]
4. Kaneko, K.; Uno, K.; Jinno, R.; Fujita, S. Prospects for phase engineering of semi-stable Ga_2O_3 semiconductor thin films using mist chemical vapor deposition. *J. Appl. Phys.* **2022**, *131*, 090902. [CrossRef]
5. Prajapati, P.; Lodha, S. Barrier height enhancement in b-Ga_2O_3 Schottky diodes using an oxygen-rich ultra-thin AlO_x interfacial layer. *Appl. Phys. Lett.* **2024**, *125*, 061602. [CrossRef]
6. Wurfl, J.; Palacios, T.; Xing, H.G.; Hao, Y.; Schubert, M. Special topic on Wide- and ultrawide-bandgap electronic semiconductor devices. *Appl. Phys. Lett.* **2024**, *125*, 070401. [CrossRef]
7. Jia, Y.; Traore, A.; Morita, R.; Florena, F.F.; Islam, M.M.; Sakurai, T. Deep-level transient spectroscopy of defect states at p-type oxide/β-Ga_2O_3 heterojunctions. *Appl. Phys. Lett.* **2024**, *125*, 073504. [CrossRef]
8. Ebrahimi-Darkhaneh, H.; Shekarnoush, M.; Arellano-Jimenez, J.; Rodriguez, R.; Colombo, L.; Quevedo-Lopez, M.; Banerjee, S.K. High-quality Mg-doped p-type Ga_2O_3 crystalline thin film by pulsed laser. *J. Mater. Sci. Mater. Electron.* **2022**, *33*, 24244–24259. [CrossRef]
9. Aleksandrova, M.; Ivanova, T.; Hamelmann, F.; Strijkova, V.; Gesheva, K. Study of Sputtered ZnO:Ga_2O_3 Films for Energy Harvesting Applications. *Coatings* **2020**, *10*, 650. [CrossRef]
10. Liu, Y.; Peng, C.; Liu, C.; Yu, C.; Guo, J.; Chang, Y.; Zhao, Y. Improved-Performance Amorphous Ga_2O_3 Photodetectors Fabricated by Capacitive Coupled Plasma-Assistant Magnetron Sputtering. *Coatings* **2024**, *14*, 1204. [CrossRef]
11. Yoon, Y.; Kim, S.; Lee, I.G.; Cho, B.J.; Hwang, W.S. Electrical and photocurrent properties of a polycrystalline Sn-doped β-Ga2O3 thin film. *Mater. Sci. Semicond. Process.* **2021**, *121*, 105430. [CrossRef]
12. Bosio, A.; Parisini, A.; Lamperti, A.; Borelli, C.; Fornasini, L.; Bosi, M.; Cora, I.; Fogarassy, Z.; Pécz, B.; Zolnai, Z.; et al. n-Type doping of ε-Ga_2O_3 epilayers by high-temperature tin diffusion. *Acta Mater.* **2021**, *210*, 116848. [CrossRef]
13. Chikoidze, E.; Sartel, C.; Yamano, H.; Zeyu Chi, Z.; Bouchez, G.; Jomard, F.; Sallet, V.; Guillot, G.; Boukheddaden, K.; Pérez-Tomás, A.; et al. Electrical properties of p-type Zn:Ga_2O_3 thin films. *J. Vac. Sci. Technol. A* **2022**, *40*, 043401. [CrossRef]
14. Chikoidze, E.; Leach, J.; Chi, Z.; Bardeleben, J.; Ballesteros, B.; Gonçalves, A.M.; Tchelidze, T.; Dumont, Y.; Pérez-Tomás, A. Surface two-dimensional hole gas in Si doped β-Ga_2O_3 thin film. *J. Alloys Compd.* **2024**, *971*, 172713. [CrossRef]
15. Lin, T.Y.; Han, S.Y.; Huang, C.Y.; Yang, C.F.; Wei, S. Using Different Methods to Measure the Optical Energy Bandgap of Un-annealed and Annealed Ga_2O_3 Films. *Appl. Funct. Mater.* **2021**, *1*, 25. [CrossRef]
16. Pech, S.; Kim, S.; Kim, N.H. Magnetron Sputter-Deposited-Ga_2O_3 Films on c-Sapphire Substrate: Effect of Rapid Thermal Annealing Temperature on Crystalline Quality. *Coatings* **2022**, *12*, 140. [CrossRef]
17. Liu, Y.; He, T.; Wei, S. Study on Preparing β-Ga_2O_3 Films with Temperature-Controlled Buffer Layer by RF Magnetron Sputtering. *Appl. Funct. Mater.* **2022**, *2*, 24. [CrossRef]
18. Correia, F.C.; Bundaleski, N.; Teodoro, O.M.N.D.; Correia, M.R.; Rebouta, L.; Mendes, A.; Tavares, C.J. XPS analysis of ZnO:Ga films deposited by magnetron sputtering: Substrate bias effect. *Appl. Surf. Sci.* **2018**, *458*, 1043–1049. [CrossRef]
19. Cui, H.; Sai, Q.; Qi, H.; Zhao, J.; Si, J.; Pan, M. Analysis on the electronic trap of b-Ga_2O_3 single crystal. *J. Mater. Sci.* **2019**, *54*, 12643–12649. [CrossRef]
20. Abenuz Acuña, J.R.; Perez, I.; Sosa, V.; Gamboa, F.; Elizalde, J.T.; Farías, R.; Carrillo, D.; Enríquez, J.L.; Burrola, A.; Mani, P. Sputtering power effects on the electrochromic properties of NiO films. *Opt. Int. J. Light Electron Opt.* **2021**, *231*, 166509. [CrossRef]
21. Salunkhe, P.; AliAV, M.; Kekuda, D. Investigation on tailoring physical properties of Nickel Oxide thin films grown by dc magnetron sputtering. *Mater. Res. Express* **2020**, *7*, 016427. [CrossRef]
22. Lin, C.L.; Wang, F.H.; Jhuang, H.S.; Yang, C.F. Effects of Different Annealing Temperatures on the Physical, Optical, and Electrical Characteristics and Chemical Bonds of Ga and F Co-Doped ZnO Films. *J. Mater. Res. Technol.* **2020**, *9*, 6331–6342. [CrossRef]

coatings

Article

First-Principles Study on the Mechanical Properties of Ni_3Sn_4-Based Intermetallic Compounds with Ce Doping

Ruisheng Zhao [1], Yan Cao [2], Jinhu He [2], Jianjun Chen [2], Shiyuan Liu [2], Zhiqiang Yang [2], Jinbao Lin [1] and Chao Chang [1,*]

[1] School of Applied Science, Taiyuan University of Science and Technology, Taiyuan 030024, China; s202318110833@stu.tyust.edu.cn (R.Z.); linjinbao@tyust.edu.cn (J.L.)
[2] Shanxi Diesel Engine Industry Co., Ltd., Datong 037036, China; 13994362433@163.com (Y.C.); weizhuanqin126@126.com (J.H.); chenjianjunde@163.com (J.C.); liushiyuan1125@163.com (S.L.); yangzhiqiang112024@163.com (Z.Y.)
* Correspondence: cc@tyust.edu.cn

Abstract: Ni_3Sn_4 intermetallic compound (IMC) is a critical material in modern electronic packaging and soldering technology. Although Ni_3Sn_4 enhances the strength of solder joints, its brittleness and anisotropy make it prone to crack formation under mechanical stress, such as thermal cycling or vibration. To improve the plasticity of Ni_3Sn_4 and mitigate its anisotropy, this study employs first-principles calculations to investigate the mechanical properties and electronic structure of the doped compounds Ce_x $Ni_{3-x}Sn_4$ (x = 0, 0.5, 1, 1.5, 2) by adding the rare earth element Ce. The results indicate that the structure $Ce_{0.5}$ $Ni_{2.5}Sn_4$ has a lower formation enthalpy (H_f) compared to other doped structures, suggesting enhanced stability. It was found that all structures exhibit improved plasticity with Ce doping, while the $Ce_{0.5}$ $Ni_{2.5}Sn_4$ structure shows relatively minor changes in hardness (H) and elastic modulus, along with the lowest anisotropy value (A^U). Analysis of the total density of states (TDOS) and partial density of states (PDOS) reveals that the electronic properties are primarily influenced by the Ni-d and Ce-f orbitals. At the Fermi level, all Ce_x $Ni_{3-x}Sn_4$ (x = 0, 0.5, 1, 1.5, 2) structures exhibit metallic characteristics and distinct electrical conductivity. Notably, the TDOS value at the Fermi level for $Ce_{0.5}$ $Ni_{2.5}Sn_4$ lies between those of Ni_3Sn_4 and other doped structures, indicating good metallicity and conductivity, as well as relative stability. Further PDOS analysis suggests that Ce doping enhances the plasticity of Ni_3Sn_4. This study provides valuable insights for the further application of rare earth elements in electronic packaging materials.

Keywords: intermetallic compounds; first-principles calculations; mechanical properties; electronic structures; rare earth elements

Academic Editor: Hideyuki Murakami

Received: 2 December 2024
Revised: 30 December 2024
Accepted: 3 January 2025
Published: 7 January 2025

Citation: Zhao, R.; Cao, Y.; He, J.; Chen, J.; Liu, S.; Yang, Z.; Lin, J.; Chang, C. First-Principles Study on the Mechanical Properties of Ni_3Sn_4-Based Intermetallic Compounds with Ce Doping. *Coatings* **2025**, *15*, 59. https://doi.org/10.3390/coatings15010059

1. Introduction

In modern electronic packaging technology, the study of intermetallic compounds (IMCs) has gained increasing importance, particularly concerning their reliability [1,2]. As solder joints become smaller and electronic devices advance towards higher performance and density integration [3,4], the reliability of IMCs is critical to the mechanical performance and durability of solder joints, directly impacting the overall performance and lifespan of electronic devices [5]. The composition of intermetallic compounds (IMCs) and strain-induced mismatch in the coefficient of thermal expansion (CTE) can lead to mechanical fatigue failures in packaging [6]. Research has shown that changes in the microstructure and mechanical properties of solder joints are often the primary causes of reliability issues [7–9].

Sn-based lead-free solder typically refers to lead-free solder with Tin (Sn) as the primary component, widely used in electronic soldering. When Tin-based solder contacts nickel-coated layers, diffusion reactions occur at high temperatures, resulting in the formation of Ni_3Sn_4 intermetallic compound [10]. This compound is of great interest due to its excellent mechanical properties, corrosion resistance, and wear resistance [11]. However, the brittleness and anisotropy of Ni_3Sn_4 can lead to cracking under thermal cycling and mechanical stress, thereby impacting the reliability of solder joints [12,13]. Therefore, enhancing the ductility of Ni_3Sn_4 and reducing its anisotropy are essential for improving the reliability of electronic packaging.

First-principles calculations, based on quantum mechanics, provide high-precision results without relying on empirical parameters or experimental data. This approach is widely utilized in the design and optimization of new materials, particularly for predicting their structures and properties [14–16]. Research by Yao et al. indicates that the incorporation of copper atoms reduces the stability and mechanical properties of the Ni_3Sn_4 phase. As the concentration of copper increases, the bulk modulus, shear modulus, and elastic modulus of $(Ni,Cu)_3Sn_4$ all decrease, suggesting that copper inclusion may adversely affect mechanical performance and increase the risk of interfacial microcracks [17,18]. Furthermore, studies by Bi and Hu et al. demonstrate that adding small amounts of Co (12.46 at.%) enhances the mechanical properties of Ni_3Sn_4 intermetallic compound, improving the shear strength of solder joints. First-principles calculations also reveal the primary reasons for this enhancement [19]. Han and Chen et al. investigated Ni_3Sn_4 with various added elements (Pd, Pt, Ge, Sb, Co, Cu, In, and Pb) and showed that copper and antimony negatively impact ductility, while germanium and lead improve it. The other elements have a minimal effect on ductility and brittleness [20]. Additionally, researchers have explored adding trace amounts of rare earth elements (such as La, Ce, and Sc) as alloying agents in solders. Studies have demonstrated that solder containing rare earth elements exhibits better wettability, creep strength, and tensile strength, while also reducing the thickness of interfacial intermetallic compounds [21,22]. For example, Zhang et al. added 0.03 wt% cerium to SnAgCu solder, and found that the fatigue life of the SnAgCuCe solder joints exceeded that of SnAgCu joints by 30.2%. This indicates that SnAgCuCe solder can replace traditional SnAgCu solder in electronic applications, offering enhanced reliability [23]. Ce was chosen for doping due to its superior ability to improve the ductility and toughness of Ni_3Sn_4, attributed to its unique electronic properties, such as d-f hybridization and increased density of states near the Fermi level. Comparative studies indicate that while other rare earth elements [24], such as La and Sc, enhance strength and hardness, their effects on plasticity and toughness are less pronounced. By introducing rare elements, the grain structure of the solder is effectively refined, which enhances the material's toughness. Furthermore, these elements improve the wettability between the solder and the substrate, thereby increasing the bonding strength of the welded joints. However, research on the doping of Ni_3Sn_4 with rare earth elements remains relatively scarce. Therefore, exploring the effects of rare earth element doping on the mechanical properties of Ni_3Sn_4 alloys, particularly concerning plasticity and anisotropy, could help to address this research gap.

This study aims to evaluate the potential of rare earth element Ce doping in Ni_3Sn_4 alloys, particularly in terms of enhancing toughness and influencing the alloy's anisotropy. By conducting an in-depth analysis of the mechanisms associated with Ce doping, this research seeks to explore new pathways for optimizing material performance, ultimately achieving superior overall mechanical properties. Initially, we examine the structural characteristics and stability of the modified compound, denoted as $Ce_xNi_{3-x}Sn_4$. Subsequently, we investigate how different levels of Ce doping influence the mechanical properties and anisotropy of the material. Finally, we analyze the electronic density of states of the

$Ce_xNi_{3-x}Sn_4$ alloy. This research aims to provide valuable insights for improving the properties of Ni_3Sn_4.

2. Structural Characteristics and Stability

The Ni_3Sn_4 crystal is classified within the monoclinic system, belonging to the space group C2/m [25]. Each unit cell comprises six nickel (Ni) atoms and eight tin (Sn) atoms. The Ni atoms occupy two distinct positions: Ni(2a) and Ni(4i). The Ni(2a) atoms are located at the eight vertices of a hexahedron and at the center of the XY plane, while the Ni(4i) atoms reside within the hexahedron and the XZ plane. The initial lattice parameters for the geometric optimization of the Ni_3Sn_4 structure are as follows: a_0 = 12.21 Å, b_0 = 4.05 Å, c_0 = 5.20 Å, β = 105.03°, and the volume V = 248.434 Å^3. A schematic diagram of the Ni_3Sn_4 crystal structure is presented in Figure 1a. Ce atom doping can substitute Ni atoms at the Wyckoff positions in the crystal, as illustrated in Figure 1. The calculated formation enthalpies for $Ce_{0.5}Ni_{2.5}Sn_4$ at the 4i and 2a positions are −24.607 kJ/mol and −23.079 kJ/mol, respectively. This indicates a preference for Ce atoms to occupy the 4i position over the 2a position, as evidenced by the lower formation enthalpy of $Ce_{0.5}Ni_{2.5}Sn_4$ (4i). Therefore, it is assumed in this calculation that all alloying elements occupy the 4i position within the Ni_3Sn_4 crystal.

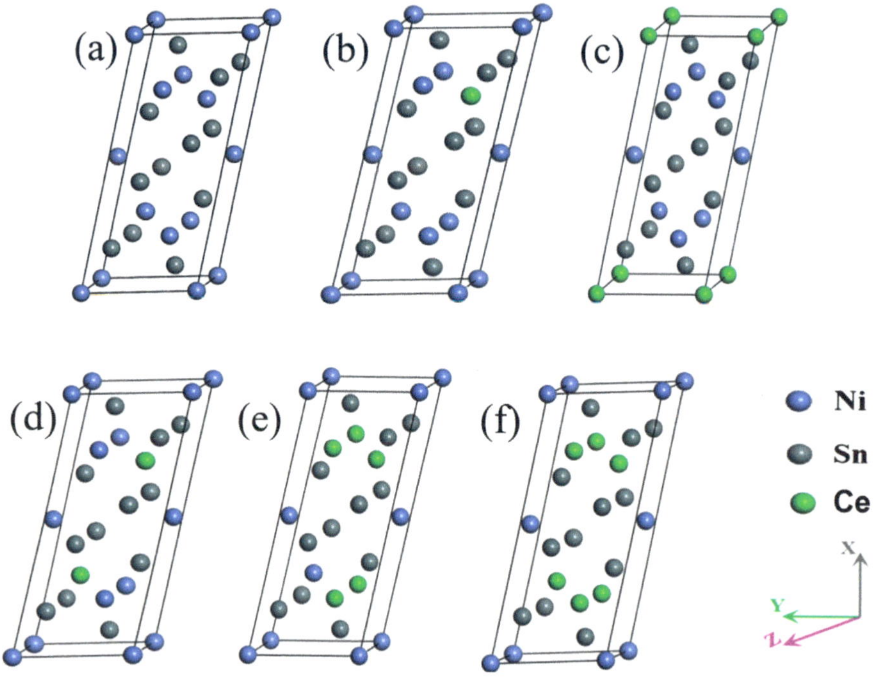

Figure 1. Crystal structures of **(a)** Ni_3Sn_4, **(b)** $Ce_{0.5}Ni_{2.5}Sn_4$(4i), **(c)** $Ce_{0.5}Ni_{2.5}Sn_4$(2a), **(d)** $Ce_1Ni_2Sn_4$, **(e)** $Ce_{1.5}Ni_{1.5}Sn_4$, and **(f)** $Ce_2Ni_1Sn_4$.

To investigate the thermodynamic stability, the following formula is employed to calculate the formation enthalpy H_f of the Ni_3Sn_4-based ternary structure [8]. This enthalpy reflects the alloy's ability to form: the more negative the H_f value, the easier it is for the alloy structure to form, indicating a more stable system [25]. the formation enthalpy is calculated using the following formula:

$$H_f = \frac{\left[E^T_{U_pV_qW_g} - (pE_U + qE_V + gE_W)\right]}{p+q+g} \quad (1)$$

where $E^T_{U_pV_qW_g}$ represents the total energy of the intermetallic compound $U_pV_qW_g$ (where U, V, and W are the constituent elements, and p, q, and g are their respective atomic numbers); E_U, E_V and E_W are the per-atom energies in the single crystal structures of U, V, and W, respectively. The calculation results are presented in Table 1. It can be observed that after Ce doping, the value of $Ce_{0.5}Ni_{2.5}Sn_4$ is relatively low and close to that of the undoped Ni_3Sn_4. Compared to other structures with varying Ce doping contents, $Ce_{0.5}Ni_{2.5}Sn_4$ exhibits greater stability.

Table 1. Through experiments and calculations, the lattice parameters (Å), cell volume (^{3}), and formation enthalpy (Hf).

Phase	Reference	a (Å)	b (Å)	c (Å)	β (○)	Volume ($\mathring{A}^3$)	Hf (kJ·mol^{-1})
Ni_3Sn_4	This work	12.210	4.054	5.196	105.03	248.434	−24.744
	Calculation [26,27]	12.220	4.060	5.270	104.970	253.46	−24.000
	Calculation [20]	12.299	4.084	5.288	105.190	-	−26.000
	Calculation [18]	12.418	4.111	4.315	105.480	-	−26.700
	Calculation [19]	12.334	4.100	5.325	105.010	-	−26.900
	Experiment [28]	12.210	4.060	5.22	105.50	258.83	-
	Experiment [29]	12.199	4.061	5.22	105.25	249.58	-
$Ce_{0.5}Ni_{2.5}Sn_4(4i)$	This work	12.283	4.093	5.322	105.41	257.941	−24.607
$Ce_{0.5}Ni_{2.5}Sn_4$(2a)	This work	12.278	4.062	5.210	105.08	250.892	−23.079
$Ce_1Ni_2Sn_4$	This work	12.493	4.114	5.231	105.32	259.300	−24.472
$Ce_{1.5}Ni_{1.5}Sn_4$	This work	12.496	4.658	5.365	105.31	301.195	−24.341
$Ce_2Ni_1Sn_4\ i$	This work	12.783	4.891	5.423	105.73	326.357	−24.214

Figure 2 illustrates the variation in lattice constants of the alloy $Ce_xNi_{3-x}Sn_4$ (x = 0, 0.5, 1, 1.5, 2). Atomic doping can significantly affect the lattice constants. Doping atoms typically either replace or insert themselves into the lattice of the matrix atoms. This alteration can lead to an increase or decrease in the lattice constant, depending on the size and properties of the doping atoms [30]. When the radius of the doping atoms is similar to that of the matrix atoms, the change in lattice constant is usually minimal. Conversely, when the radius of the doping atoms exceeds that of the matrix atoms, the lattice constant tends to increase; if the doping atoms are smaller, the lattice constant decreases. As the amount of Ce replacing Ni increases, the lengths of a (Å), b (Å), and c (Å) exhibit different increases compared to the lattice constant of Ni_3Sn_4, showing distinct directional differences. When the increments of the lattice constants a (Å), b (Å), and c (Å) differ, the crystal typically exhibits noticeable anisotropy in various directions [31,32]. This may weaken the directional dependence of bonding and consequently impact the anisotropic and elastic properties of $Ce_xNi_{3-x}Sn_4$ (x = 0, 0.5, 1, 1.5, 2), which will be discussed later.

In this study, first-principles calculations based on density functional theory (DFT) were performed using the Cambridge Sequential Total Energy Package (CASTEP) [33]. Traditional nanoindentation allows for rapid and practical assessment of material mechanical properties [34–36], while first-principles calculations provide high precision and a deep understanding of material characteristics. During the calculations, the electron exchange-correlation function was modeled using the Perdew–Burke–Ernzerhof (PBE) [37], functional under the generalized gradient approximation (GGA-PBE) [38], and the interaction potential between valence electrons and ions was simulated using the ultrasoft pseudopotential (USPP). A kinetic energy cutoff of 400 eV was selected for the plane wave function expansion, and the Broyden–Fletcher–Goldfarb (BFG) algorithm was employed for structural optimization. The Pulay density mixing method was utilized for electron relaxation in the self-consistent field (SCF) calculations. The convergence criteria for the SCF calculations were set as follows: the total energy must be less than 1.0×10^{-6} eV/atom, the force on each atom must be less than 0.05 eV/Å, the tolerance offset must be less than 0.002 Å, and the maximum stress deviation must be less than 0.1 GPa.

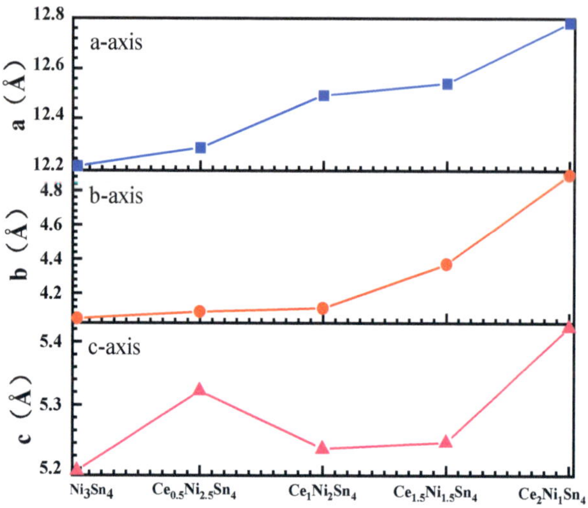

Figure 2. Changes in lattice constants a (Å), b (Å), and c (Å) with varying Ce content.

To ensure the accuracy of subsequent calculations and obtain the most stable crystal structure, a stringent total energy convergence test was conducted. The elastic constants C_{ij} and elastic compliance constants S_{ij} for $Ce_xNi_{3-x}Sn_4$ were calculated using $2 \times 5 \times 6$ Monkhorst–Pack K points, and the results are presented in Tables 2 and 3.

Table 2. Elastic constants (GPa) of $Ce_xNi_{3-x}Sn_4$ (x = 0, 0.5, 1, 1.5, 2).

Phase	Reference	C_{11}	C_{12}	C_{13}	C_{22}	C_{23}	C_{33}	C_{44}	C_{55}	C_{66}	C_{15}	C_{25}	C_{35}	C_{46}
Ni_3Sn_4	This work	181.06	68.70	60.12	146.08	75.24	162.32	62.74	60.61	47.03	−24.46	11.23	−11.34	14.92
	Calculation [20]	176.60	64.30	57.40	146.00	76.10	169.70	60.80	49.50	46.70	−22.90	7.00	−9.6	6.90
	Calculation [18]	155.48	70.68	69.34	164.33	68.26	149.86	62.74	59.99	59.95	−21.97	13.99	−8.73	4.90
$Ce_{0.5}Ni_{2.5}Sn_4$	This work	168.12	65.37	60.08	143.16	65.21	157.03	55.11	53.51	43.32	−17.6	12.42	−9.71	11.21
$Ce_1Ni_2Sn_4$	This work	152.87	63.28	59.21	122.79	64.52	140.27	52.66	52.91	43.67	−16.88	11.23	−11.34	9.73
$Ce_{1.5}Ni_{1.5}Sn_4$	This work	140.13	55.79	50.12	119.17	73.30	137.71	51.64	50.49	43.21	−16.02	10.07	−15.12	9.27
$Ce_2Ni_1Sn_4$	This work	125.31	45.61	40.27	93.22	63.57	104.28	51.13	40.53	32.01	−6.11	7.01	−11.09	4.38

Table 3. Elastic compliance constants (GPa) of $Ce_xNi_{3-x}Sn_4$ (x = 0, 0.5, 1, 1.5, 2).

Phase	S_{11}	S_{12}	S_{13}	S_{22}	S_{23}	S_{33}	S_{44}	S_{55}	S_{66}	S_{15}	S_{25}	S_{35}	S_{46}
Ni_3Sn_4	0.0076	−0.0033	−0.0010	0.0108	−0.0041	0.0086	0.0172	0.0190	0.0232	0.0035	−0.0041	0.0019	−0.0055
$Ce_{0.5}Ni_{2.5}Sn_4$	0.0081	−0.0033	−0.0016	0.0103	−0.0033	0.0084	0.0192	0.0210	0.0244	0.0031	−0.0041	0.0018	−0.0050
$Ce_1Ni_2Sn_4$	0.0093	−0.0040	−0.0018	0.0126	−0.0045	0.0102	0.0198	0.0216	0.0239	0.0034	−0.0049	0.0026	−0.0044
$Ce_{1.5}Ni_{1.5}Sn_4$	0.0095	−0.0043	−0.0008	0.0156	−0.0075	0.0121	0.0201	0.0238	0.0241	0.0036	−0.0067	0.0049	−0.0043
$Ce_2Ni_1Sn_4$	0.0100	−0.0044	−0.0010	0.0220	−0.0126	0.0184	0.0198	0.0283	0.0316	0.0020	−0.0079	0.0071	−0.0027

3. Mechanical Properties

The mechanical stability of a crystal structure can be assessed using independent elastic constants [39]. The elastic constant matrix can be reduced to 13 variables based on the symmetry of the monoclinic structure. The criteria for mechanical stability in the monoclinic system are expressed by the following equations [40–42]:

$$C_{ij} > 0(i = 1 - 6) \tag{2}$$

$$[C_{11} + C_{22} + C_{33} + 2(C_{12} + C_{13} + C_{23})] > 0 \tag{3}$$

$$(C_{33}C_{55} - C_{35}) > 0, (C_{44}C_{66} - C_{46}) > 0, (C_{22} + C_{33} - 2C_{23}) > 0 \tag{4}$$

Table 2 presents the calculated elastic constants C_{ij} for all structures based on Ni_3Sn_4. The elastic constants for Ni_3Sn_4 closely align with other computational results [18,20]. The elastic constants C_{ij} for the structures of $Ce_xNi_{3-x}Sn_4$ (x = 0, 0.5, 1, 1.5, 2) readily satisfy the corresponding criteria listed in Equations (2)–(4), indicating that the studied $Ce_xNi_{3-x}Sn_4$ structures are mechanically stable at 0 K. To further compute the Young's modulus [43] of $Ce_xNi_{3-x}Sn_4$ (x = 0, 0.5, 1, 1.5, 2), the Voigt–Reuss–Hill (VRH) [44] approach can be employed. The Voigt (V) and Reuss (R) values for bulk modulus and shear modulus can be calculated using the following formulas:

$$B_V = \left(\frac{1}{9}\right)[C_{11} + C_{22} + C_{33} + 2(C_{12} + C_{13} + C_{23})] \tag{5}$$

$$\begin{aligned} G_V = (1/15)[C_{11} + C_{22} + C_{33} + 3(C_{44} + C_{55} + C_{66}) \\ -(C_{12} + C_{13} + C_{23})] \end{aligned} \tag{6}$$

$$\begin{aligned} B_R = \Omega[a(C_{11} + C_{22} - 2C_{12}) + b(2C_{12} - 2C_{11} - C_{23}) \\ +c(C_{15} - 2C_{25}) + d(2C_{12} + 2C_{23} - C_{13} - 2C_{22}) \\ +2e(C_{25} - C_{15}) + f]^{-1} \end{aligned} \tag{7}$$

$$\begin{aligned} G_R = 15\{4[a(C_{11} + C_{22} + C_{12}) + b(C_{11} - C_{12} - C_{23}) + c(C_{15} \\ +C_{25}) + d(C_{22} - C_{12} - C_{23} - C_{13}) + e(C_{15} - C_{25}) + f]/\Omega \\ +3\left[\frac{g}{\Omega} + \frac{C_{44}+C_{66}}{C_{44}C_{66}-C_{46}^2}\right]\}^{-1} \end{aligned} \tag{8}$$

$$a = C_{33}C_{55} - C_{35}^2 \tag{9}$$

$$b = C_{23}C_{55} - C_{25}C_{35} \tag{10}$$

$$c = C_{13}C_{35} - C_{15}C_{33} \tag{11}$$

$$d = C_{13}C_{55} - C_{15}C_{35} \tag{12}$$

$$e = C_{13}C_{25} - C_{15}C_{23} \tag{13}$$

$$\begin{aligned} f = C_{11}(C_{22}C_{55} - C_{25}^2) - C_{12}(C_{12}C_{55} - C_{15}C_{25}) \\ +C_{15}(C_{12}C_{25} - C_{15}C_{22}) + C_{25}(C_{23}C_{35} - C_{25}C_{33}) \end{aligned} \tag{14}$$

$$g = C_{11}C_{22}C_{33} - C_{11}C_{23}^2 - C_{22}C_{13}^2 - C_{33}C_{12}^2 + 2C_{12}C_{13}C_{23} \tag{15}$$

$$\Omega = 2\begin{bmatrix} C_{15}C_{25}(C_{33}C_{12} - C_{13}C_{23}) + C_{15}C_{35}(C_{22}C_{13} - C_{12}C_{23}) \\ +C_{25}C_{35}(C_{11}C_{23} - C_{12}C_{13}) \end{bmatrix} \\ -[C_{15}^2(C_{22}C_{33} - C_{23}^2) + C_{25}^2(C_{11}C_{33} - C_{13}^2) + C_{35}^2(C_{11}C_{22} - C_{12}^2)] + gC_{55} \tag{16}$$

The bulk modulus (B) represents the resistance of a material to external compression in flexible systems, while the shear modulus (G) describes the material's ability to resist shear strain. The Young's modulus (E) and Poisson's ratio (ν) can be derived from G and B. Additionally, hardness (H) can be obtained from the Young's modulus (E) and Poisson's ratio (ν) [45,46], with calculations as follows:

$$B = \frac{B_V + B_R}{2} \tag{17}$$

$$G = \frac{G_V + G_R}{2} \tag{18}$$

$$E = \frac{9BG}{3B + G} \tag{19}$$

$$v = \frac{3B - 2G}{2(3B + G)} \tag{20}$$

$$H = \frac{(1 - 2v)}{6(1 + v)}E \tag{21}$$

The values of the bulk modulus (B), shear modulus (G), Young's modulus (E), Poisson's ratio (ν), hardness (H), and the ratio B/G for $Ce_xNi_{3-x}Sn_4$ ($x = 0, 0.5, 1, 1.5, 2$) can be derived from Equations (17) to (21). The calculated values are presented in Figure 3 and Table 4. Regarding the hardness of pure Ni_3Sn_4, some experimental data are listed in Table 5. A large number of nanoindentation results for Young's modulus fall within the range of 119.4 to 156 GPa, while the hardness values range from 6.1 to 8.8 GPa. The main variation in these values is due to the influence of the sample's specific morphology on the nanoindentation measurement process. In this study, for pure Ni_3Sn_4, the calculated values are as follows: B = 99.15 GPa, G = 49.59 GPa, E = 128.31 GPa, ν = 0.284, and B/G = 1.98. These values fall within the range of experimental data obtained via nanoindentation, further confirming the reliability of the computational approach [19,47]. The capacity of a material to withstand deformation is represented by its bulk modulus (B), determined by the strength of the chemical bonds and the compressibility of the material. A higher bulk modulus signifies increased strength, which corresponds to a diminished ability to deform. Table 4 indicates that the bulk modulus (B) decreases with higher Ce doping levels, implying that Ce doping enhances the plasticity of the material.

Table 4. Calculated elastic modulus and relevant values of $Ce_x\ Ni_{3-x}Sn_4$ alloys (unit in GPa).

Phase	Reference	B_V	G_V	B_R	G_R	B	G	E	v	H	B/G
Ni_3Sn_4	This work	99.15	53.03	98.58	46.87	99.15	49.95	128.31	0.284	7.18	1.98
	Calculation [48]	-	-	-	-	100.86	49.89	128.48	0.287	-	2.02
	Calculation [47]	-	-	-	-	100.19	51.62	132.15	0.280	-	1.94
	Calculation [18]	98.47	53.96	97.56	49.42	98.01	53.96	136.78	0.270	-	1.82
$Ce_{0.5}Ni_{2.5}Sn_4$	This work	94.50	48.87	93.94	44.98	94.22	46.92	120.72	0.286	6.68	2.00
$Ce_1Ni_2Sn_4$	This work	88.21	45.38	87.49	41.02	87.85	43.19	111.34	0.289	6.08	2.03
$Ce_{1.5}Ni_{1.5}Sn_4$	This work	83.89	43.56	82.75	37.22	83.32	40.39	104.32	0.291	5.61	2.06
$Ce_2Ni_1Sn_4$	This work	69.07	36.29	68.77	29.27	68.92	32.78	84.88	0.295	4.49	2.10

Table 5. The Young's modulus (*E*) and hardness of Ni_3Sn_4 were obtained through experiments and calculations. (unit in GPa).

Phase	Reference	*E*	*H*
Ni_3Sn_4	This work	128.31	7.18
Ni_3Sn_4	Experiment [19]	124.6	5.81
Ni_3Sn_4	Experiment [49]	123.4~134.0	6.1~7.0
Ni_3Sn_4	Experiment [18]	135.3 ± 8.4	5.0 ± 0.63
$(Co, Ni)_3Sn_4$	Experiment [19]	132.3	5.98
$(Cu, Ni)_3Sn_4$	Experiment [18]	126.3 ± 7.6	4.7 ± 0.72

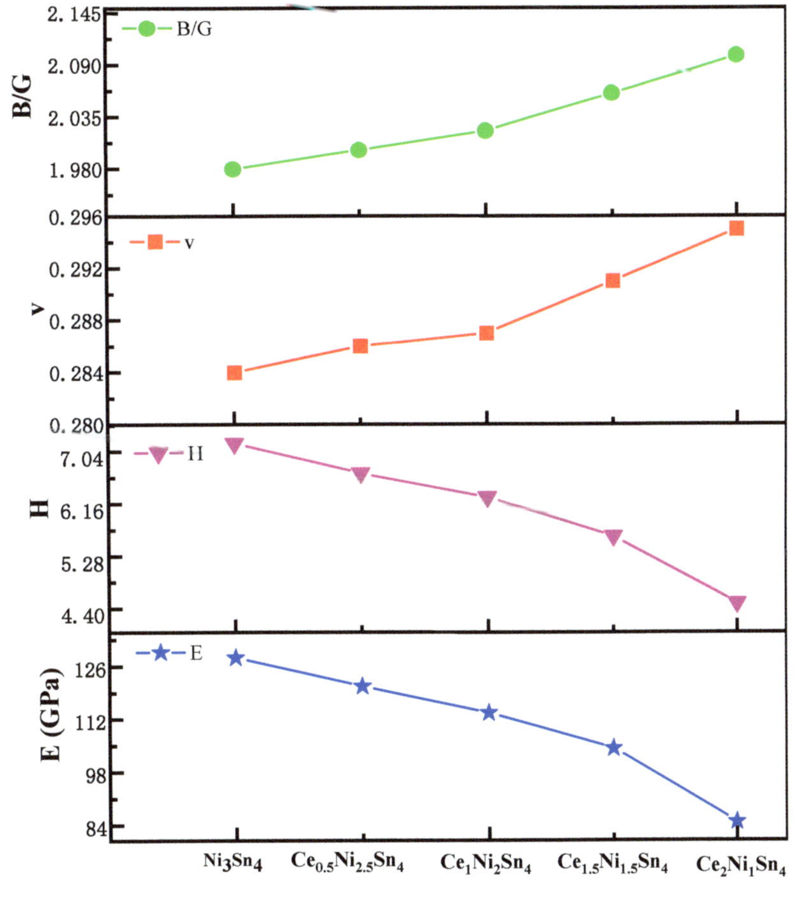

Figure 3. The variation in bulk modulus, shear modulus, and Young modulus with the Ce atom.

In general, the ratio of bulk modulus to shear modulus (*B/G*) is used to describe the ductility or brittleness of a crystal structure [46]. A lower *B/G* value is associated with brittleness, while a higher *B/G* value indicates improved ductility. For the Ni_3Sn_4 structure, the *B/G* ratio is 1.98. Following the addition of Ce, the *B/G* value increases with higher Ce content, suggesting an enhancement in ductility. Moreover, a lower Poisson's ratio signifies that the material experiences less lateral deformation during tensile or compressive loading, making it more susceptible to brittle fracture rather than plastic deformation once a critical stress threshold is reached [50]. In contrast, a higher Poisson's ratio indicates greater lateral deformation under similar loading conditions, reflecting improved deformability and greater plasticity. Thus, the Poisson's ratio is directly related to the material's ability to resist shear stress; a higher Poisson's ratio corresponds to enhanced plasticity. Consequently, the observed increase in Poisson's ratio with Ce doping suggests that the plasticity of Ni_3Sn_4 is enhanced, which is consistent with the findings from the *B/G* analysis. This indicates that with Ce doping, the material is more likely to undergo plastic deformation rather than brittle fracture under applied stress. Furthermore, as Ce content increases, the hardness (H) value decreases accordingly.

It is generally accepted that the evolution of cracks within solder joints is closely related to the elastic anisotropy of intermetallic compounds (IMCs). In electronic packaging, microcracks induced by elastic anisotropy can adversely affect the reliability of electronic products. During the production and operation of solder joints, the anisotropic plastic deformation of IMCs can lead to the formation of microcracks, thereby reducing the reliability of electronic devices [51]. To investigate the impact of Ce introduction on the elastic anisotropy of IMCs, a universal anisotropy index A^U was employed. A higher A^U value indicates greater structural anisotropy of the material. The calculation formula for A^U is provided below, with the calculated results shown in Figure 4.

$$A^U = 5\frac{G_V}{G_R} + \frac{B_V}{B_R} - 6 \tag{22}$$

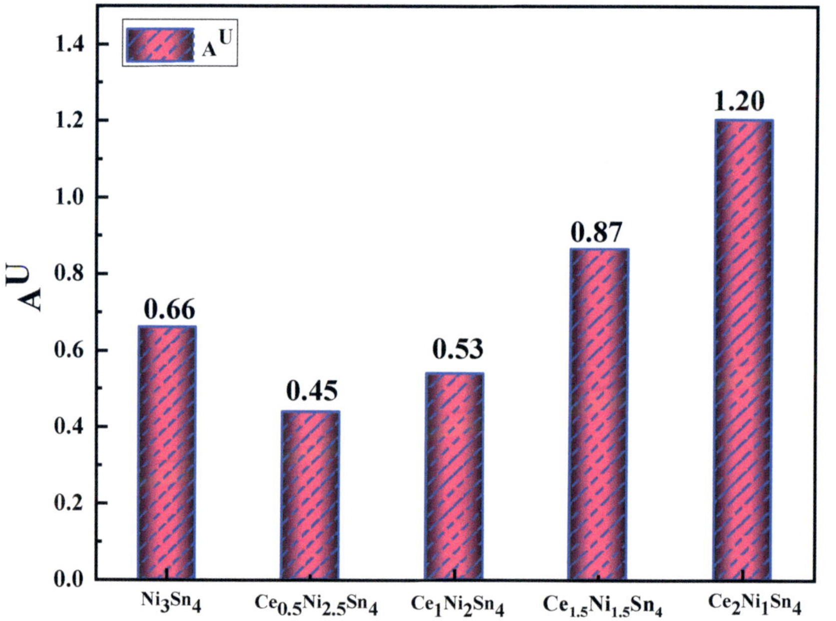

Figure 4. The variation of A^U with the Ce atom fraction.

If the value of A^U is zero, the structure is considered isotropic. A higher A^U value indicates greater anisotropy in the material. The A^U value for Ni_3Sn_4 is 0.66, which is close to values reported in the literature [19]. Calculations suggest that $Ce_x\ Ni_{3-x}Sn_4$ is anisotropic, with $Ce_{0.5}Ni_{2.5}Sn_4$ exhibiting a smaller A^U value than Ni_3Sn_4. This indicates that a small amount of Ce doping can improve the anisotropy of Ni_3Sn_4, thereby reducing the tendency for microcrack formation, which helps enhance the reliability of the solder joints.

To further investigate the anisotropy of the $Ce_x\ Ni_{3-x}Sn_4$ structure, three-dimensional (3D) surface plots of the bulk modulus and Young's modulus were created, as shown in Figures 5 and 6. The results indicate that for all Ni_3Sn_4 based structures, the three-dimensional surface plots display varying degrees of deviation from sphericity, suggesting that both Ni_3Sn_4 and $Ce_x\ Ni_{3-x}Sn_4$ exhibit anisotropy. Furthermore, after Ce doping, the shape of $Ce_x\ Ni_{3-x}Sn_4$ deviates even more from sphericity, indicating an enhancement in anisotropy. Notably, only $Ce_{0.5}Ni_{2.5}Sn_4$ exhibits a shape closer to spherical for its Young's modulus, suggesting that this structure has relatively lower anisotropy, which is consistent with the computed results.

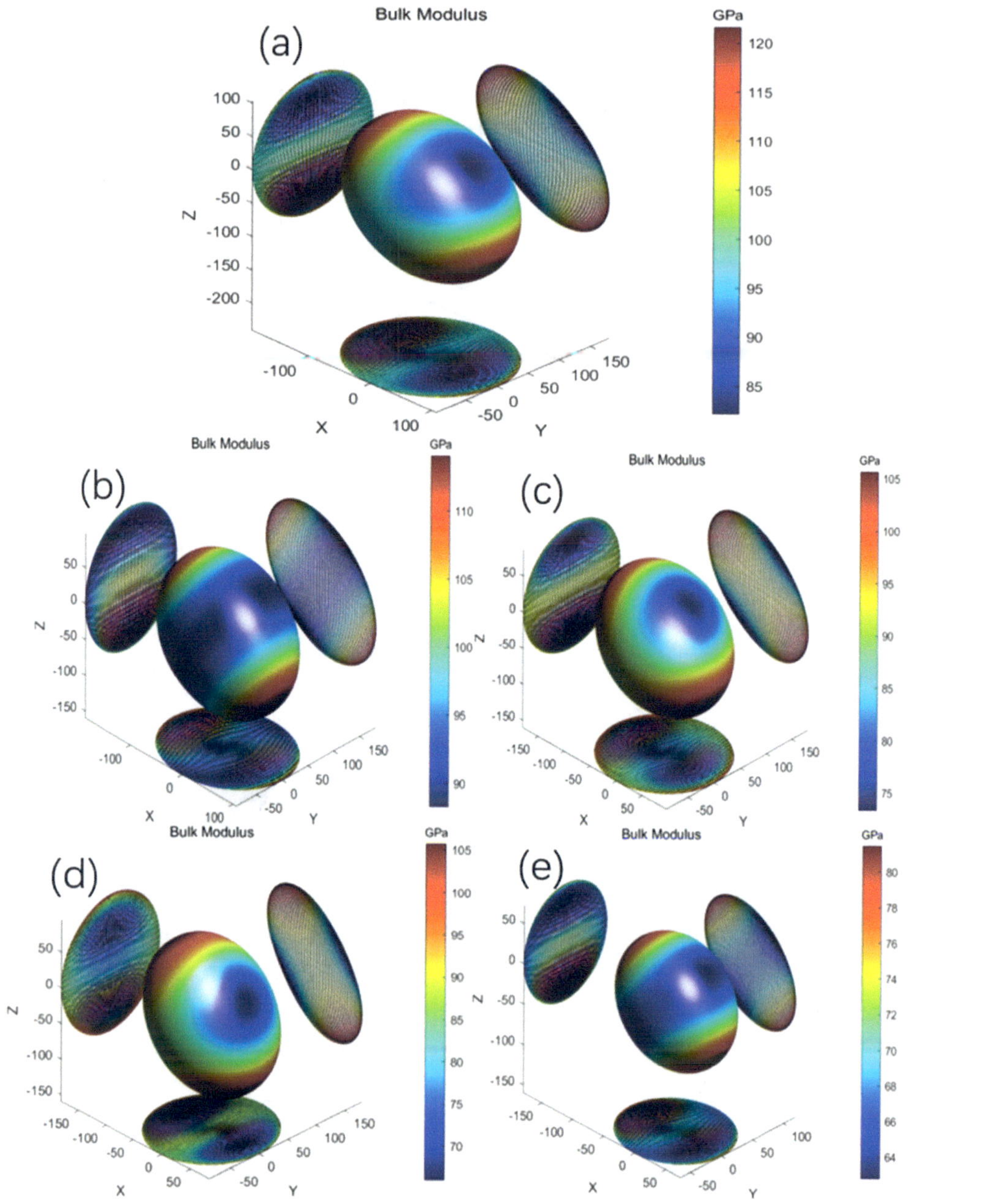

Figure 5. The surface constructions of bulk modulus for (**a**) Ni_3Sn_4, (**b**) $Ce_{0.5}Ni_{2.5}Sn_4$, (**c**) $Ce_1Ni_2Sn_4$, (**d**) $Ce_{1.5}Ni_{1.5}Sn_4$, and (**e**) $Ce_2Ni_1Sn_4$.

In summary, the plasticity of the $Ce_x\ Ni_{3-x}Sn_4$ structures is enhanced compared to the undoped version. Notably, the $Ce_{0.5}Ni_{2.5}Sn_4$ structure not only demonstrates improved plasticity but also exhibits the lowest degree of anisotropy, which reduces the tendency for microcrack formation. This enhancement contributes to the reliability of the solder joints.

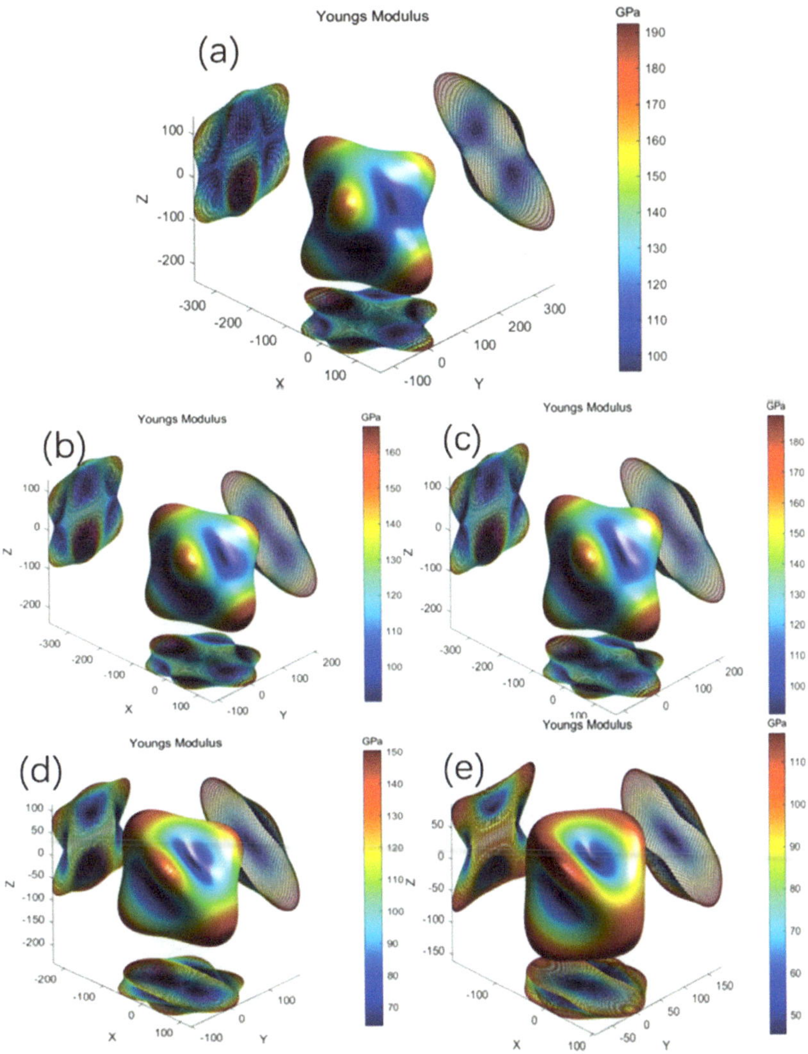

Figure 6. The 3D surface plots of Young's modulus for (**a**) Ni_3Sn_4, (**b**) $Ce_{0.5}Ni_{2.5}Sn_4$, (**c**) $Ce_1Ni_2Sn_4$, (**d**) $Ce_{1.5}Ni_{1.5}Sn_4$, and (**e**) $Ce_2Ni_1Sn_4$.

4. Electronic Structure

It is widely recognized that the mechanical properties and energy characteristics of materials are largely determined by their electronic structure. To better understand the bonding mechanisms of intermetallic compounds with different alloying elements, we analyzed the total density of states (TDOS) and partial density of states (PDOS) for $Ce_xNi_{3-x}Sn_4$ (x = 0, 0.5, 1, 1.5, 2). The results are illustrated in Figures 7 and 8, where the Fermi level (E_F) is positioned at zero energy. For all studied structures, the alloy density of states curves are continuous at the Fermi energy (0 eV), and the density of states (DOS) values at the Fermi level (N_{EF}) are all greater than zero, indicating that all structures exhibit metallic characteristics and good electrical conductivity [52]. A smaller density of states at the Fermi level correlates with greater structural stability; conversely, a larger density can lead to bonding instability within the system [53]. Figure 7 shows that the alloy density of states at the Fermi level increases with higher Ce content. Therefore, as the Ce content increases, the structural stability decreases gradually. This result is consistent with the conclusion from the previous formation energy calculations, which show that as the Ce content increases, the system's energy increases, leading to a gradual decline in structural stability. Among the $Ce_xNi_{3-x}Sn_4$ structures, $Ce_{0.5}Ni_{2.5}Sn_4$ has the lowest alloy density of states at the Fermi level, consistent with previous structural stability analyses. Furthermore, it was found that the density of states at the Fermi level (0 eV) for the doped structures is higher than that of Ni_3Sn_4, suggesting an enhancement in conductivity following doping. A greater electronic density of states within the alloy system facilitates the formation of metallic bonds, which contributes to improved ductility and reduced brittleness [53]. The continued addition of Ce is expected to enhance the alloy's ductility, aligning with the results derived from *B/G*, Poisson's ratio ν, and Cauchy pressure C_{ij} assessments regarding the impact of Ce content on the alloy's ductility-brittleness balance.

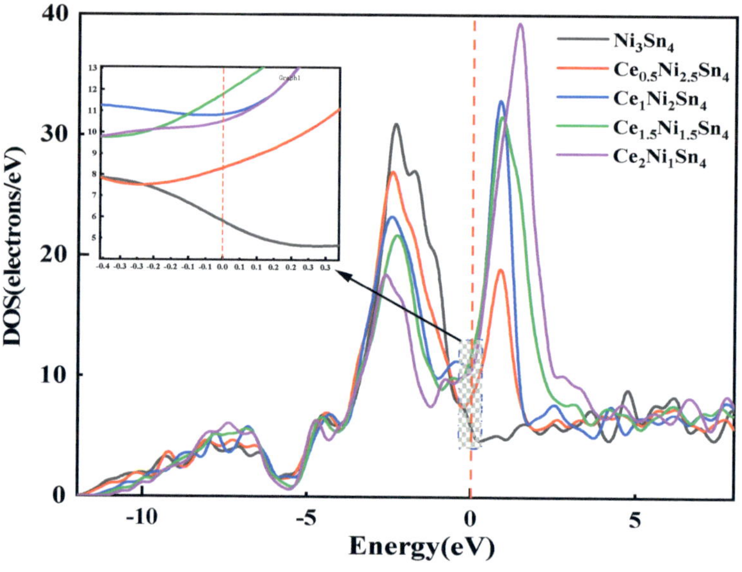

Figure 7. Comparison of total density of states (TDOS) in $Ce_xNi_{3-x}Sn_4$ (x = 0, 0.5, 1, 1.5, 2).

In metal compound Ni_3Sn_4, two primary bonding peaks are observed at −2.31 eV and −1.78 eV. The main factor influencing the bonding in Ni_3Sn_4 is the hybridization between Ni-d and Sn-p states, with the DOS values contributed by Ni-d states spanning from 0 to −5.5 eV. Below −6 eV, the Sn-s states predominantly influence the DOS, consistent with previous findings [54]. In the Ce-doped intermetallic compound $Ce_xNi_{3-x}Sn_4$, these bonding peaks are weakened, and a new bonding peak appears around 0.8–1.4 eV, primarily dominated by Ce-f electrons. This indicates that the hybridization between Ni-d and Sn-p electrons is diminished after Ce replaces Ni in Ni_3Sn_4. The introduction of Ce also reveals a contribution from Ce-f states to the DOS and suggests that there may be covalent bonding between Ni and Ce.

Furthermore, the formation of covalent interactions aligns with the trends observed in the Poisson's ratio. In the Ce-doped Ni_3Sn_4 system studied here, we observed a significant increase in the overlap between Ce-f and Sn-p orbitals with increasing Ce content. This indicates that upon Ce substitution of Ni_3Sn_4, the hybridization between Ni-d and Sn-p electrons is weakened. Additionally, a new peak appears in the 0 eV to 4 eV range, primarily dominated by Ce-d electrons. The strength of atomic interactions can be inferred from the overlap of f- and p-hybrid orbitals. According to Gschneidner et al. [55], a clear relationship between electronic structure and the ductile or brittle behavior of crystal structures has been established. Therefore, we hypothesize that Ce doping enhances the ductility of Ni_3Sn_4, which is consistent with the increased ductility observed in our mechanical property calculations.

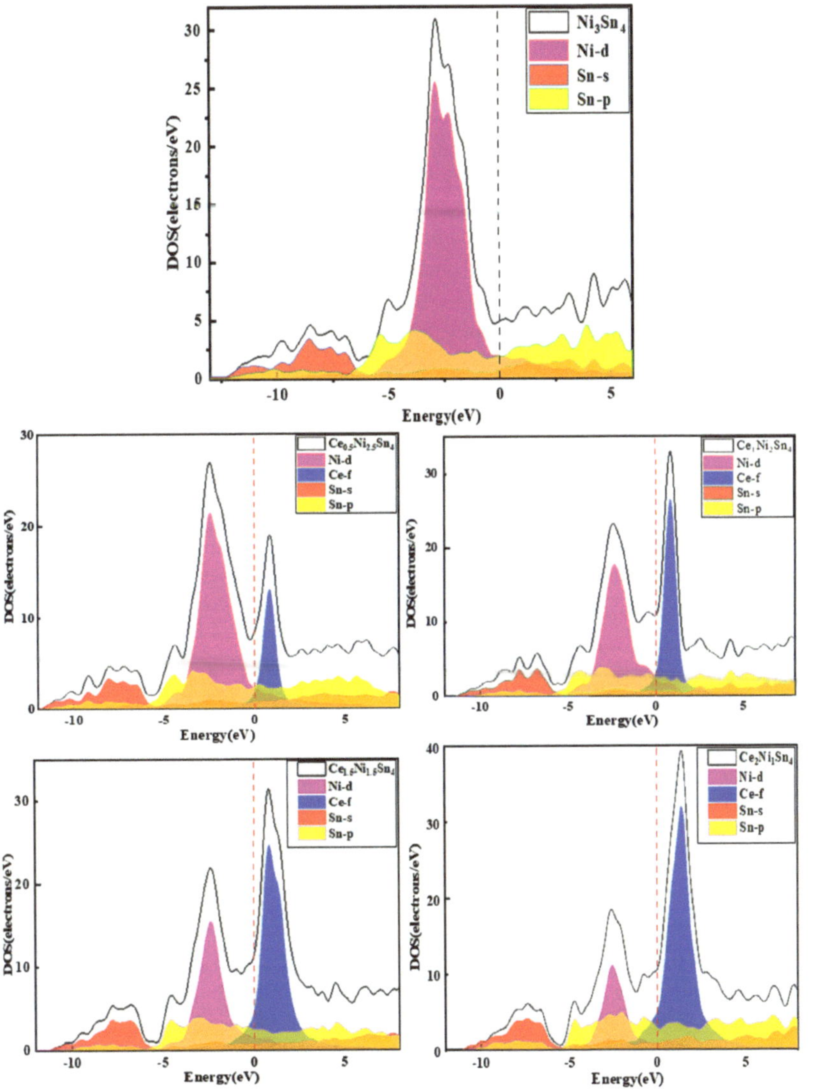

Figure 8. Total and partial density of states (TPDOS).

In summary, the electronic structure analysis reveals that the total density of states (TDOS) and partial density of states (PDOS) of $Ce_xNi_{3-x}Sn_4$ (x = 0, 0.5, 1, 1.5, 2) are primarily influenced by Ni-d and Ce-f states. The addition of Ce increases the toughness of the Ni_3Sn_4 alloy, aligning with the trends observed in the previous calculations.

5. Conclusions

In this study, we employed first-principles calculations to simulate the ternary intermetallic compound based on Ni_3Sn_4 and investigate the structural, elastic, and electronic properties following the substitution of Ni with cerium (Ce). The results can be summarized as follows:

(1) In the doped structure of $Ce_{0.5}Ni_{2.5}Sn_4$, the formation enthalpies for Ce occupying the 4i and 2a sites are −24.607 kJ/mol and −23.079 kJ/mol, respectively. This indicates that Ce atoms preferentially occupy the 4i site over the 2a site. Additionally, the formation enthalpy H_f value for the structure with x = 0.5 is the lowest among the $Ce_xNi_{3-x}Sn_4$ (x = 0, 0.5, 1, 1.5, 2) structures, suggesting that Ce doping favors the stability of the $Ce_{0.5}Ni_{2.5}Sn_4$(4i) structure relative to the others.

(2) After Ce doping, the elastic modulus of the $Ce_x Ni_{3-x}Sn_4$ structure decreased, while its plasticity improved across the board. Additionally, both the Poisson's ratio (ν) and the ratio of bulk modulus to shear modulus (*B/G*) increased with higher Ce doping levels. Notably, the $Ce_{0.5}Ni_{2.5}Sn_4$ structure exhibited minimal changes in hardness (*H*) and elastic modulus, yet its anisotropy value (A^U) was the lowest, indicating that Ce doping effectively enhances the ductility and reduces the anisotropic characteristics of Ni_3Sn_4.

(3) The electronic structure analysis of $Ce_xNi_{3-x}Sn_4$ (x = 0, 0.5, 1, 1.5, 2) reveals that the total density of states (TDOS) and partial density of states (PDOS) are primarily influenced by Ni-d and Ce-f orbitals. All studied structures exhibit metallic characteristics and good electrical conductivity, as evidenced by nonzero TDOS at the Fermi level. Increasing Ce content leads to higher TDOS at the Fermi level, indicating reduced structural stability, consistent with formation energy calculations. The introduction of Ce also weakens the hybridization between Ni-d and Sn-p states, while generating a new bonding peak at 0.8~1.4 eV dominated by Ce-f orbitals. This suggests the formation of covalent bonds between Ni and Ce, and an enhanced overlap between Ce-f and Sn-p orbitals with higher Ce content. These electronic changes contribute to stronger metallic bonding and improved ductility, aligning with mechanical property calculations, such as *B/G* ratios, Poisson's ratio, and Cauchy pressure. Overall, the analysis confirms that Ce doping enhances the toughness of Ni_3Sn_4, providing insights into the role of electronic structure in tailoring the mechanical properties of intermetallic compounds.

To date, there have been no reported experimental or theoretical values. Therefore, the results of this study can provide a reference for the further application of this type of material and serve as a validation for future research efforts.

Author Contributions: R.Z. (First Author): Conceptualization, methodology, investigation, simulation, writing—original draft; J.C.: investigation; J.H.: investigation; Y.C.: investigation; Z.Y.: methodology, investigation; J.L.: methodology; S.L.: conceptualization; C.C. (corresponding author): conceptualization, funding acquisition, resources, supervision, writing—review and editing. All authors have read and agreed to the published version of the manuscript.

Funding: This work was supported by The Regional Collaboration Project of Shanxi Province (No. 202204041101044).

Institutional Review Board Statement: Not applicable.

Informed Consent Statement: Not applicable.

Data Availability Statement: All data generated or analyzed during this study are included in this published article.

Conflicts of Interest: Yan Cao, Jinhu He, Jianjun Chen, Shiyuan Liu and Zhiqiang Yang were employed by the company Shanxi Diesel Engine Industry Co., Ltd. The remaining authors declare that the research was conducted in the absence of any commercial or financial relationships that could be construed as a potential conflict of interest.

References

1. Chen, J.; Yang, J.; Zhang, Y.; Yu, Z.; Zhang, P. Effect of Substrates on the Formation of Kirkendall Voids in Sn/Cu Joints. *Weld. World* **2019**, *63*, 751–757. [CrossRef]
2. Han, Y.; Yang, J.; Xu, L.; Jing, H.; Zhao, L. Effect of Transient Current Bonding on Interfacial Reaction in Ag-Coated Graphene Sn–Ag–Cu Composite Solder Joints. *Trans. Nonferrous Met. Soc. China* **2021**, *31*, 2454–2467. [CrossRef]
3. Cao, K.; Zhou, J.; Wei, T.; Chen, M.; Hu, S.; Li, K. A Survey of Optimization Techniques for Thermal-Aware 3D Processors. *J. Syst. Archit.* **2019**, *97*, 397–415. [CrossRef]
4. Le, F.; Lee, S.-W.R.; Zhang, Q. 3D Chip Stacking with through Silicon-Vias (TSVs) for Vertical Interconnect and Underfill Dispensing. *J. Micromech. Microeng.* **2017**, *27*, 045012. [CrossRef]
5. Zhang, Z.; Chen, J.; Wang, J.; Han, Y.; Yu, Z.; Wang, Q.; Zhang, P.; Yang, S. Effects of Solder Thickness on Interface Behavior and Nanoindentation Characteristics in Cu/Sn/Cu Microbumps. *Weld. World* **2022**, *66*, 973–983. [CrossRef]
6. Long, X.; Su, T.; Chen, Z.; Su, Y.; Siow, K.S. Tunable Coefficient of Thermal Expansion of Composite Materials for Thin-Film Coatings. *Coatings* **2022**, *12*, 836. [CrossRef]
7. Li, X.; Wang, L.; Li, X.; Li, C.; Sun, F.; Liu, M.; Zhu, L.; Li, M. The Effects of Solder Joint Height on the Solid–Liquid Interface Diffusion in Micro Solder Joints. *Mater. Lett.* **2022**, *316*, 131969. [CrossRef]
8. Chen, J.; Zhang, H.; Zhang, P.; Yu, Z.; Zhang, Y.; Yu, C.; Lu, H. The Zn Accumulation Behavior, Phase Evolution and Void Formation in Sn-xZn/Cu Systems by Considering Trace Zn: A Combined Experimental and Theoretical Study. *J. Mater. Res. Technol.* **2019**, *8*, 4141–4150. [CrossRef]
9. Zou, H.F.; Zhang, Q.K.; Zhang, Z.F. Interfacial Microstructure and Mechanical Properties of SnBi/Cu Joints by Alloying Cu Substrate. *Mater. Sci. Eng. A* **2012**, *532*, 167–177. [CrossRef]
10. Yang, Y.-S.; Yang, C.-J.; Ouyang, F.-Y. Interfacial Reaction of Ni_3Sn_4 Intermetallic Compound in Ni/SnAg Solder/Ni System under Thermomigration. *J. Alloys Compd.* **2016**, *674*, 331–340. [CrossRef]
11. Rafailović, L.D.; Karnthaler, H.P.; Trišović, T.; Minić, D.M. Microstructure and Mechanical Properties of Disperse Ni–Co Alloys Electrodeposited on Cu Substrates. *Mater. Chem. Phys.* **2010**, *120*, 409–416. [CrossRef]
12. Tu, K. Reliability Challenges in 3D IC Packaging Technology. *Microelectron. Reliab.* **2011**, *51*, 517–523. [CrossRef]
13. Shen, J.; Chan, Y.; Liu, S. Growth Mechanism of Ni_3Sn_4 in a Sn/Ni Liquid/Solid Interfacial Reaction. *Acta Mater.* **2009**, *57*, 5196–5206. [CrossRef]
14. Chen, J.-S.; Ye, C.-H.; Yu, C.; Lu, H. The Micro-Mechanism for the Effect of Sn Grain Orientation on Substrate Consumption in Sn Solder Joints. *Comput. Mater. Sci.* **2015**, *108*, 1–7. [CrossRef]
15. Yang, A.; Xiao, K.; Duan, Y.; Li, C.; Yi, J.; Peng, M.; Shen, L. Effect of Indium on Microstructure, Mechanical Properties, Phase Stability and Atomic Diffusion of Sn-0.7Cu Solder: Experiments and First-Principles Calculations. *Mater. Sci. Eng. A* **2022**, *855*, 143938. [CrossRef]
16. Mattes, S.; Brennan, S.; Woodcox, M. First Principles Modeling of Mechanical Properties of Binary Alloys Containing Ga, Sn, and In for Soldering Applications. *J. Phys. Condens. Matter* **2023**, *35*, 485401. [CrossRef] [PubMed]
17. Yang, W.; Yamamoto, T.; Aso, K.; Somidin, F.; Nogita, K.; Matsumura, S. Atom Locations in a Ni Doped η-(Cu,Ni)6Sn5 Intermetallic Compound. *Scr. Mater.* **2019**, *158*, 1–5. [CrossRef]
18. Bi, X.; Hu, X.; Jiang, X.; Li, Q. Effect of Cu Additions on Mechanical Properties of Ni3Sn4-Based Intermetallic Compounds: First-Principles Calculations and Nano-Indentation Measurements. *Vacuum* **2019**, *164*, 7–14. [CrossRef]
19. Bi, X.; Hu, X.; Li, Q. Effect of Co Addition into Ni Film on Shear Strength of Solder/Ni/Cu System: Experimental and Theoretical Investigations. *Mater. Sci. Eng. A* **2020**, *788*, 139589. [CrossRef]
20. Han, Y.; Chen, J.; Lin, M.; Zhang, K.; Lu, H. Synergistic Effects of Alloy Elements on the Structural Stability, Mechanical Properties and Electronic Structure of Ni3Sn4: Using First Principles. *Vacuum* **2023**, *214*, 112239. [CrossRef]
21. Ma, X.; Li, L.; Zhang, Z.; Wang, H.; Wang, E.; Qiu, T. Effects of Rare Earth La on Microstructure and Properties of Ag–21Cu–25Sn Alloy Ribbon Prepared by Melt Spinning. *Mater. Des.* **2015**, *83*, 1–5. [CrossRef]
22. Zhang, L.; Fan, X.; Guo, Y.; He, C. Properties Enhancement of SnAgCu Solders Containing Rare Earth Yb. *Mater. Des.* **2014**, *57*, 646–651. [CrossRef]
23. Zhang, L.; Han, J.; Guo, Y.; He, C. Effect of Rare Earth Ce on the Fatigue Life of SnAgCu Solder Joints in WLCSP Device Using FEM and Experiments. *Mater. Sci. Eng. A* **2014**, *597*, 219–224. [CrossRef]
24. Yu, X.; Lin, G.; Zhang, Z.; Xie, J. Electronic Structure Characteristics of Fe-6.5 wt%Si Alloy Doped with Rare Earth Elements and Its Effect on Mechanical Properties. *J. Alloys Compd.* **2020**, *843*, 155916. [CrossRef]

25. He, Q.; Yu, B.; Li, Z.; Zhao, Y. Density Functional Theory for Battery Materials. *Energy Environ. Mater.* **2019**, *2*, 264–279. [CrossRef]
26. Yu, C.; Chen, J.; Xu, J.; Chen, J.; Lu, H. Effects of Additive Elements on Structural and Electronic Properties of Sn-Based Intermetallics by First Principles. *Mater. Res. Express* **2014**, *1*, 025702. [CrossRef]
27. Flandorfer, H.; Saeed, U.; Luef, C.; Sabbar, A.; Ipser, H. Interfaces in Lead-Free Solder Alloys: Enthalpy of Formation of Binary Ag–Sn, Cu–Sn and Ni–Sn Intermetallic Compounds. *Thermochim. Acta* **2007**, *459*, 34–39. [CrossRef]
28. Jeitschko, W.; Jaberg, B. Structure Refinement of Ni_3Sn_4. *Acta Cryst. B* **1982**, *38*, 598–600. [CrossRef]
29. Xian, J.W.; Zeng, G.; Belyakov, S.A.; Gu, Q.; Nogita, K.; Gourlay, C.M. Anisotropic Thermal Expansion of Ni_3Sn_4, Ag_3Sn, Cu_3Sn, Cu_6Sn_5 and βSn. *Intermetallics* **2017**, *91*, 50–64. [CrossRef]
30. Zhang, H.; Wang, N.; Wang, S.; Zhang, Y. Effect of Doping 3d Transition Metal (Fe, Co, and Ni) on the Electronic, Magnetic and Optical Properties of Pentagonal ZnO_2 Monolayer. *Phys. E Low-Dimens. Syst. Nanostruct.* **2020**, *117*, 113806. [CrossRef]
31. Bouamama, K.H.; Djemia, P.; Faurie, D.; Abadias, G. Structural and Elastic Properties of Single-Crystal and Polycrystalline Ti1−xZrxN Alloys: A Computational Study. *J. Alloys Compd.* **2012**, *536*, S138–S142. [CrossRef]
32. Jin, N.; Yan, Z.; Wang, Y.; Cheng, H.; Zhang, H. Effects of Heat Treatment on Microstructure and Mechanical Properties of Selective Laser Melted Ti-6Al-4V Lattice Materials. *Int. J. Mech. Sci.* **2021**, *190*, 106042. [CrossRef]
33. Segall, M.D.; Lindan, P.J.D.; Probert, M.J.; Pickard, C.J.; Hasnip, P.J.; Clark, S.J.; Payne, M.C. First-Principles Simulation: Ideas, Illustrations and the CASTEP Code—IOPscience. Available online: https://iopscience.iop.org/article/10.1088/0953-8984/14/11/301 (accessed on 28 June 2024).
34. Long, X.; Li, J.; Shen, Z.; Su, Y. Dimensionless Analysis to Determine Elastoplastic Properties of Thin Films by Indentation. *Coatings* **2022**, *12*, 1768. [CrossRef]
35. Long, X.; Lu, C.; Shen, Z.; Su, Y. Identification of Mechanical Properties of Thin-Film Elastoplastic Materials by Machine Learning. *Acta Mech. Solida Sin.* **2023**, *36*, 13–21. [CrossRef]
36. Long, X.; Li, Y.; Shen, Z.; Su, Y.; Gu, T.; Siow, K.S. Review of uniqueness challenge in inverse analysis of nanoindentation. *J. Manuf. Process.* **2024**, *131*, 1897–1916. [CrossRef]
37. Perdew, J.P. Generalized Gradient Approximation Made Simple. *Phys. Rev. Lett.* **1996**, *77*, 3865–3868. [CrossRef]
38. Payne, M.C. Iterative Minimization Techniques for Ab Initio Total-Energy Calculations: Molecular Dynamics and Conjugate Gradients. *Rev. Mod. Phys.* **1992**, *64*, 1045–1097. [CrossRef]
39. Boas, W. Physical Properties of Crystals, Their Representation by Tensors and Matrices: J.F Nye: Clarendon Press, Oxford, 1957. 322pp., 50s. *J. Mech. Phys. Solids* **1958**, *6*, 328–329. [CrossRef]
40. Goumri-Said, S.; Kanoun, M. Theoretical investigations of structural, elastic, electronic and thermal properties of Damiaoite $PtIn_2$. *Comput. Mater. Sci.* **2008**, *43*, 243–250. [CrossRef]
41. Beckstein, O.; Klepeis, J.E.; Hart, G.L.W.; Pankratov, O. First-Principles Elastic Constants and Electronic Structure of α-Pt_2Si and c. *Phys. Rev. B* **2001**, *63*, 134112. [CrossRef]
42. Noheda, B.; Gonzalo, J.A.; Cross, L.E.; Guo, R.; Park, S.-E.; Cox, D.E.; Shirane, G. Tetragonal-to-Monoclinic Phase Transition in a Ferroelectric Perovskite: The Structure of $PbZr_{0.52}Ti_{0.48}O_3$. *Phys. Rev. B* **2000**, *61*, 8687–8695. [CrossRef]
43. Cheng, H.-C.; Yu, C.-F.; Chen, W.-H. First-Principles Density Functional Calculation of Mechanical, Thermodynamic and Electronic Properties of CuIn and Cu_2In Crystals. *J. Alloys Compd.* **2013**, *546*, 286–295. [CrossRef]
44. Hill, R. The Elastic Behaviour of a Crystalline Aggregate. *Proc. Phys. Soc. A* **1952**, *65*, 349. [CrossRef]
45. Korozlu, N.; Colakoglu, K.; Deligoz, E.; Aydin, S. The Elastic and Mechanical Properties of MB_{12} (M = Zr, Hf, Y, Lu) as a Function of Pressure. *J. Alloys Compd.* **2013**, *546*, 157–164. [CrossRef]
46. Pugh, S.F. XCII. Relations between the Elastic Moduli and the Plastic Properties of Polycrystalline Pure Metals. *Lond. Edinb. Dublin Philos. Mag. J. Sci.* **1954**, *45*, 823–843. [CrossRef]
47. Gao, F.; Qu, J. Elastic Moduli of (Ni,Cu)3Sn4 Ternary Alloys from First-Principles Calculations. *J. Electron. Mater.* **2010**, *39*, 2429–2434. [CrossRef]
48. Zhang, W.W.; Ma, Y.; Zhou, W.; Wu, P. The Structural, Elastic and Electronic Properties of $Ni_{3-x}Cu_xSn_4$ (x = 0, 0.5, 1 and 1.5) Intermetallic Compounds via Ab Initio Calculations. *J. Electron. Mater.* **2019**, *48*, 4533–4543. [CrossRef]
49. Chen, Z.; He, M.; Balakrisnan, B.; Chum, C.C. Elasticity Modulus, Hardness and Fracture Toughness of Ni_3Sn_4 Intermetallic Thin Films. *Mater. Sci. Eng. A* **2006**, *423*, 107–110. [CrossRef]
50. Wang, Y.; Yang, J.; Huang, J.; Wang, W.; Ye, Z.; Chen, S.; Zhao, Y. First-Principles Calculations on Physical Properties of Ni_3Sn_x Binary System Intermetallic Compounds and Ni/Ni_3Sn Interfaces in Nickel-Tin TLPS Bonding Layer. *Intermetallics* **2018**, *101*, 27–38. [CrossRef]
51. Yang, P.-F.; Lai, Y.-S.; Jian, S.-R.; Chen, J.; Chen, R.-S. Nanoindentation Identifications of Mechanical Properties of Cu_6Sn_5, Cu_3Sn, and Ni_3Sn_4 Intermetallic Compounds Derived by Diffusion Couples. *Mater. Sci. Eng. A* **2008**, *485*, 305–310. [CrossRef]
52. Tian, J.; Zhao, Y.; Wen, Z.; Hou, H.; Han, P. Physical Properties and Debye Temperature of Al_7Cu_2Fe Alloy under Various Pressures Analyzed by First-Principles. *Solid State Commun.* **2017**, *257*, 6–10. [CrossRef]

..; Dai, W.; Luo, J. A Review on the Interfacial Intermetallic Compounds between Sn–Ag–Cu ...ater. Sci. Mater. Electron. **2010**, *21*, 421–440. [CrossRef]
...tive Stability of LI_2, DO_{22}, and DO_{23} Structures in MAl_3 Compounds. *J. Mater. Res.* **1989**, *4*,
...n, M.; Wang, C.Z.; Ho, K.M.; Russell, A.M.; Mudryk, Y.; Becker, A.T.; Larson, J.L. Influence of the Electronic ... the Ductile Behavior of B2 CsCl-Type AB Intermetallics. *Acta Mater.* **2009**, *57*, 5876–5881. [CrossRef]

coatings

Article

Microstructure and Mechanical Properties of TixNbMoTaW Refractory High-Entropy Alloy for Bolt Coating Applications

Ruisheng Zhao [1], Yan Cao [2], Jinhu He [2], Jianjun Chen [2], Shiyuan Liu [2], Zhiqiang Yang [2], Jinbao Lin [1] and Chao Chang [1,*]

1 School of Applied Science, Taiyuan University of Science and Technology, Taiyuan 030024, China; s202318110833@stu.tyust.edu.cn (R.Z.); linjinbao@tyust.edu.cn (J.L.)
2 Shanxi Diesel Engine Industry Co., Ltd., Datong 037036, China; 13994362433@163.com (Y.C.); weizhuanqin126@126.com (J.H.); chenjianjunde@163.com (J.C.); liushiyuan1125@163.com (S.L.); yangzhiqiang112024@163.com (Z.Y.)
* Correspondence: cc@tyust.edu.cn

Abstract: High-strength bolts are prone to crack initiation from the threaded hole during fastening due to large loads, which can compromise their performance and reliability. To enhance the durability of these bolts, coatings are often employed to strengthen their surfaces. NbMoTaW refractory high-entropy alloy coatings are widely used in hard coating applications due to their exceptional mechanical properties. However, the brittleness of this alloy at room temperature limits its performance in high-stress environments. To enhance the ductility of NbMoTaW alloys, this study systematically investigates the effect of varying titanium (Ti) content on the alloy's properties. First-principles calculations were employed to analyze the elastic properties of TixNbMoTaW alloys, including elastic constants, the elastic modulus, the bulk modulus (B)-to-shear modulus (G) ratio (Pugh's ratio), Poisson's ratio (ν), and Cauchy pressure (C12–C44). The results indicate that the addition of Ti significantly improves the alloy's plasticity. Specifically, when the Ti content is x = 2, the B/G ratio increases to 3.23, and Poisson's ratio increases to 0.39, indicating enhanced deformability. At x = 0.75, the elastic modulus (E) increases to 273.78 GPa, compared to 244.99 GPa for the original alloy. The experimental results further validate the computational findings. X-ray diffraction (XRD) and scanning electron microscopy (SEM) analyses indicate that all alloys exhibit a single body-centered cubic (BCC) phase. Room-temperature compression tests show that as the Ti content increases, the yield strength, fracture strength, and plasticity of the alloys significantly improve. Specifically, for a Ti content of x = 0.75, the yield strength reaches 1551 MPa, the fracture strength is 1856 MPa, and the plastic strain increases to 14.6%. For Ti1.5NbMoTaW, the yield strength is 1506 MPa, the fracture strength is 1893 MPa, and the plastic strain is 17.3%. Overall, TixNbMoTaW refractory high-entropy alloys demonstrate significant improvements in both plasticity and strength, showing great potential for coating applications in high-stress environments.

Keywords: refractory high-entropy alloy; phase structure; mechanical properties; first-principles calculations

Academic Editor: Raul Arrabal

Received: 20 December 2024
Revised: 13 January 2025
Accepted: 20 January 2025
Published: 21 January 2025

Citation: Zhao, R.; Cao, Y.; He, J.; Chen, J.; Liu, S.; Yang, Z.; Lin, J.; Chang, C. Microstructure and Mechanical Properties of TixNbMoTaW Refractory High-Entropy Alloy for Bolt Coating Applications. *Coatings* **2025**, *15*, 120. https://doi.org/10.3390/coatings15020120

1. Introduction

High-strength bolts experience significant loads during connection, especially under tightening loads, leading to crack initiation from the threaded holes. These cracks can not only cause bolt failure but also compromise the safety and reliability of the entire connection system. To enhance bolts' performance and extend their service life, researchers commonly apply coatings to improve the mechanical properties of the bolt materials [1–3]. As the demand for high-performance coating materials continues to grow, refractory high-entropy alloys (RHEAs) have emerged as promising candidates due to their exceptional high-temperature strength, corrosion resistance, and hardness [4–8]. RHEAs typically consist of five or more metallic elements [9–11], forming solid solutions with high structural stability. These alloys exhibit excellent mechanical properties and superior high-temperature characteristics, enabling them to offer outstanding wear resistance, oxidation resistance, and thermal corrosion resistance in extreme environments [12–15]. Despite its excellent performance at elevated temperatures, NbMoTaW RHEA exhibits significant brittleness at room temperature [16,17] which limits its broader application in certain engineering fields. To overcome this challenge, improving the room-temperature plasticity of NbMoTaW refractory high-entropy alloys has become a key focus of research, aiming to expand their applicability in a broader range of fields.

Optimizing the mechanical properties of refractory high-entropy alloys (RHEAs) to achieve higher strength and toughness has become a prominent area of research in materials science. In recent years, numerous studies have focused on doping various elements to modulate the microstructure and mechanical properties of RHEAs. Senkov et al. [18] experimentally investigated the microstructure, hardness, tensile properties, and high-temperature mechanical performance of NbMoTaW and VNbMoTaW alloys, analyzing the effect of vanadium (V) doping on alloy performance. Their findings indicated that while the addition of vanadium enhanced the high-temperature strength, it sacrificed the alloy's plasticity and toughness. Similarly, Qiao et al. [19] studied NbTaV-based alloys, improving their high-temperature mechanical properties by adding titanium (Ti) and tungsten (W). Their experiments showed that the incorporation of Ti and W significantly increased the high-temperature strength and stability, but excessive tungsten content could lead to increased brittleness at room temperature. Other reports have suggested that precise control of alloy composition and microstructure can improve overall performance, with a focus on the design of novel alloy systems [20]; Chan et al. [21] also demonstrated that in Nb alloys, sufficient Ti content (>24%) is required to achieve ductility. Ti enhances the tensile ductility and fracture toughness of Nb solid solution alloys, further confirming the positive role of Ti in enhancing the plasticity of the alloys. In addition to RHEAs, other coating strategies have also demonstrated significant potential [22]. For instance, incorporating Y_2O_3 into an Al_2O_3 matrix has been shown to enhance wear resistance and mechanical strength, although its adaptability to high-temperature and humid environments requires further optimization [23]. Similarly, high-entropy alloys, with their unique multi-principal element compositions and tunable properties, offer tremendous potential for corrosion resistance in harsh conditions [24]. Studies suggest that adding elements like titanium (Ti) [25] or niobium (Nb) can further improve the corrosion resistance and stability of HEAs, enabling them to perform reliably in complex environments. Qi et al. [26] demonstrated, through first-principles calculations, that Mo and W, as brittle metals, result in lower toughness and ductility in NbMoTaW RHEAs. Alloying to reduce the average valence electron concentration of Mo- and W-based alloys can improve their toughness and ductility. The valence electron concentrations of Ti, Nb, Mo, Ta, and W are 4, 5, 6, 5, and 6, respectively. Since Ti has the lowest valence electron concentration, introducing Ti into NbMoTaW alloys reduces the overall valence electron concentration, thereby enhancing their toughness. Therefore, the addition of Ti holds significant research importance for enhancing the plasticity, toughness, and ductility of NbMoTaW refractory high-entropy alloys (RHEAs).

This study systematically investigates the effects of titanium (Ti) content on the microstructure and room-temperature mechanical properties of TixNbMoTaW (x = 0, 0.25, 0.5, 0.75, 1, 1.5, 2) refractory high-entropy alloys (RHEAs), based on NbMoTaW alloys, using a combination of first-principles calculations and experimental testing. First, first-principles calculations were employed to analyze the influence of varying Ti content on the mechanical properties and electronic structure of TixNbMoTaW alloys, focusing on changes in elastic constants, elastic moduli, B/G ratio, Poisson's ratio (ν), Cauchy pressure (C12–C44), anisotropy factor, and electronic density of states. Subsequently, the alloy samples were characterized using X-ray diffraction (XRD) and scanning electron microscopy (SEM). Finally, room-temperature compression tests were performed to further validate the effectiveness of Ti in enhancing the alloy's plasticity and to confirm the reliability of the first-principles calculations in predicting the mechanical properties of the alloys.

2. Computational Theory and Experimental Methods

2.1. Theoretical Methods

In this study, the Special Quasirandom Structures (SQS) method is used for modeling. Figure 1 shows the unit cell model of the body-centered cubic (BCC) structure for the TixNbMoTaW refractory high-entropy alloy. In this study, first-principles calculations based on density functional theory (DFT) were performed using the Cambridge Sequential Total Energy Package (CASTEP) [27]. Traditional nanoindentation allows for the rapid and practical assessment of material mechanical properties [28–30], while first-principles calculations provide high precision and a deep understanding of material characteristics. During the calculations, the electron exchange-correlation function was modeled using the Perdew–Burke–Ernzerhof (PBE) [31] functional under the Generalized Gradient Approximation (GGA-PBE) [32], the interaction potential between valence electrons and ions was simulated using the ultrasoft pseudopotential (USPP), and the plane-wave function expansion cutoff energy was set to 650 eV. Geometry optimization of the structure was carried out using the Broyden–Fletcher–Goldfarb (BFG) algorithm. For the Self-Consistent Field (SCF) calculations, electron relaxation was performed using the Pulay density mixing method. The convergence criteria for the SCF calculations were as follows: the total energy was converged to within 1.0×10^{-6} eV/atom, the force on each atom was less than 0.05 eV/Å, the displacement tolerance was below 0.002 Å, and the maximum stress deviation was smaller than 0.1 GPa. To ensure the accuracy of the computational results, the optimal K-point grid was selected based on the lowest energy criterion. As shown in Figure 2, the energy is lowest and the convergence is best when the K-point grid is set to $10 \times 10 \times 10$ for all five alloys. Therefore, in this study, a K-point grid of $10 \times 10 \times 10$ was used for the calculations.

2.2. Experimental Details

In this study, TixNbMoTaW refractory high-entropy alloys were melted using a vacuum arc furnace (AM200) from Edmund Bühler, Bodelshausen, Germany. First, the alloy components were precisely calculated and weighed according to the designed composition (see Table 1). The metal raw materials were polished to remove oxide layers using sandpaper, then cleaned with alcohol and dried, as shown in Figure 3, to ensure a purity of over 99.95%, with particle sizes controlled between 1 and 10 mm. Prior to melting, the equipment was inspected to ensure proper operation, and the furnace chamber was evacuated to eliminate air and prevent reactions at high temperatures. During the melting process, argon gas was used as the protective atmosphere. The melting was conducted at a voltage of 380 V and in a current range of 250–500 A, with the arc temperature maintained between 5000 and 6000 K. Each melting cycle lasted for 5 min, and the ingot was flipped nine times to ensure homogeneity in the alloy composition.

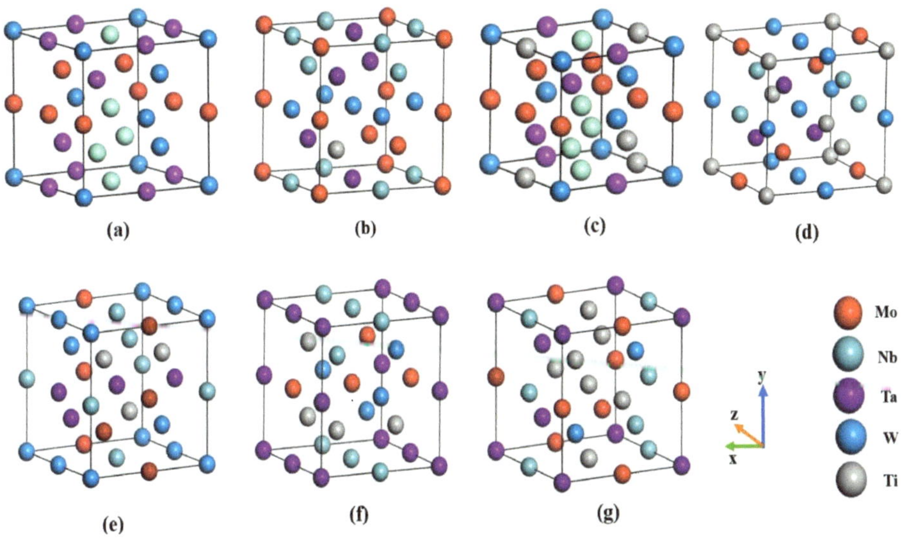

Figure 1. Cell model of $Ti_xNbMoTaW$ high-entropy alloy: (**a**) NbMoTaW, (**b**) $Ti_{0.25}NbMoTaW$, (**c**) $Ti_{0.5}NbMoTaW$, (**d**) $Ti_{0.75}NbMoTaW$, (**e**) TiNbMoTaW, (**f**) $Ti_{1.5}NbMoTaW$, (**g**) $Ti_2NbMoTaW$.

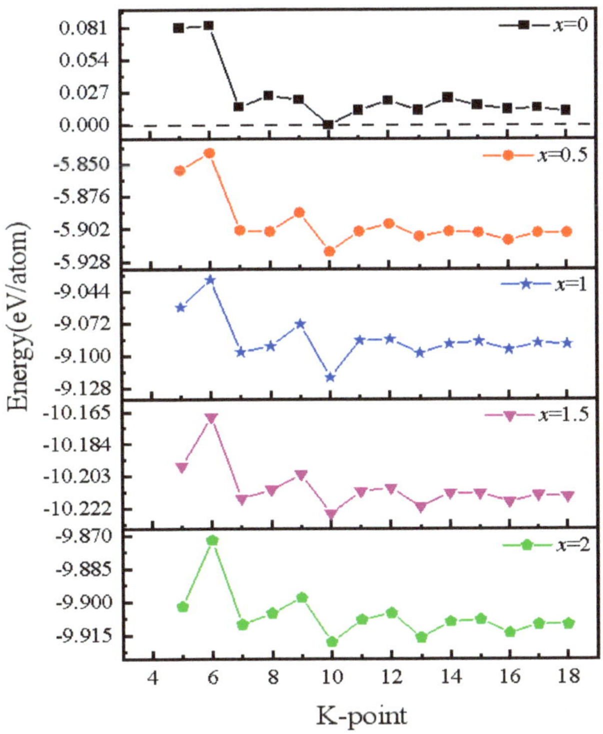

Figure 2. Convergence curve of k-point grid and total energy of $Ti_xNbMoTaW$ high-entropy alloys.

Table 1. Percentage of elements in TixNbMoTaW refractory high-entropy alloy (in%).

	Ti	Nb	Mo	Ta	W
NbMoTaW	0	25.00	25.00	25.00	25.00
$Ti_{0.25}$NbMoTaW	5.88	23.53	23.53	23.53	23.53
$Ti_{0.5}$NbMoTaW	11.11	22.22	22.22	22.22	22.22
$Ti_{0.75}$NbMoTaW	15.79	21.05	21.05	21.05	21.05
TiNbMoTaW	20.00	20.00	20.00	20.00	20.00
Ti1.5NbMoTaW	27.27	18.18	18.18	18.18	18.18
Ti2NbMoTaW	33.33	16.66	16.66	16.66	16.66

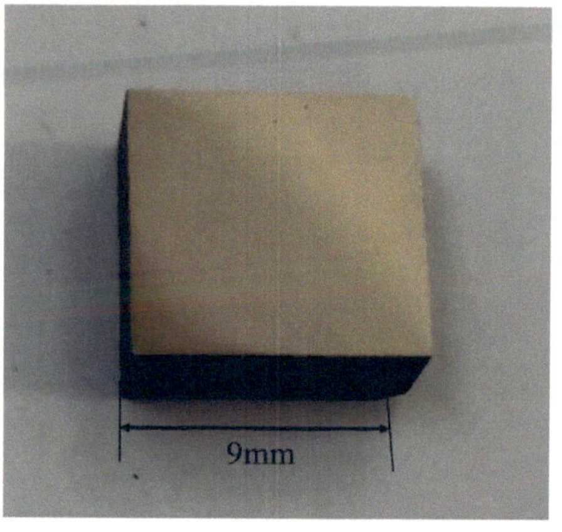

Figure 3. TixNbMoTaW refractory high-entropy alloy characterization sample.

In this experiment, the TixNbMoTaW refractory high-entropy alloy was characterized using X-ray diffraction (XRD) with a SmartLab-SE X-ray diffractometer from Rigaku, Tokyo, Japan. Additionally, the microstructure and morphology of the cast alloy were analyzed using a TESCAN MIRA LMS scanning electron microscope (SEM) from the Brno, Czech Republic, equipped with an energy-dispersive spectrometer (EDS) and backscattered electron detector (BSE) to evaluate the degree of element segregation and the elemental distribution in the microregions. XRD analysis was used to determine the phase structure of the alloys, providing insights into their crystallographic properties. The XRD patterns were analyzed based on the diffraction angles (2θ) and the corresponding peak intensities, allowing for the identification of the crystalline phases present in the samples. Additionally, room-temperature compression tests were conducted on the alloy samples, which were shaped as cylinders with a length-to-diameter ratio of 1.5:1, and dimensions of $\varphi 3 \times 4.5$ mm. The selected strain rate (loading rate) for the uniaxial compression test was 5×10^{-4} s^{-1}.

3. Results and Discussion

3.1. Theoretical Results

3.1.1. Crystal Structure

This study uses computational analysis to determine the specific phase types of a high-entropy alloy based on parameters such as atomic radius difference (δ), mixing enthalpy (ΔH_{mix}), mixing entropy (ΔS_{mix}), entropy-to-enthalpy ratio (Ω), electronegativity difference ($\Delta \chi$), atomic size difference parameter (γ), and average valence electron concentration (VEC) [33,34]. The atomic radius difference (δ) reflects the disparity in atomic radii among the constituent elements. The entropy-to-enthalpy ratio (Ω) indicates the relative weight of entropy and enthalpy in the alloy, helping to assess the alloy's entropy level and preventing the formation of intermetallic compounds. When the alloy parameters meet the following conditions, the alloy exhibits a disordered solid solution phase: δ < 6.5%, $-15 \leq \Delta H_{mix} \leq 5$ kJ/mol, $12 \leq \Delta S_{mix} \leq 17.5$ J/(K·mol), and $\Omega \geq 1.1$ [29–32]. An atomic size difference parameter (γ) < 1.175 and a small electronegativity difference indicate a tendency for the alloy to form a solid solution. When the average valence electron concentration (VEC) is ≤6.87, the alloy crystallizes in a body-centered cubic (BCC) structure; for $6.87 \leq \text{VEC} \leq 8$, the alloy adopts a dual-phase BCC and face-centered cubic (FCC) structure, and when VEC ≥ 8, the alloy forms an FCC structure [35].

The calculated phase formation parameters for the TixNbMoTaW refractory high-entropy alloy are shown in Figure 4. The phase structure of TixNbMoTaW high-entropy alloys is found to be a disordered solid solution phase. The calculated VEC values are less than 6.87, indicating a BCC structure, and the VEC decreases from 5.500 to 4.998, suggesting that the addition of Ti improves the alloy's toughness and ductility. Figure 4 illustrates the impact of Ti content on the alloy phase formation parameters. As the Ti content increases, δ, ΔS_{mix}, Ω, and γ initially increase and then decrease. The maximum δ of 1.891 occurs at x = 1.5, while the maximum Ω of 5.801 is observed at x = 0.5. Meanwhile, the mixing enthalpy (ΔH_{mix}), electronegativity difference (Δχ), and VEC decrease significantly, further confirming that TixNbMoTaW high-entropy alloys are more likely to form a BCC structure.

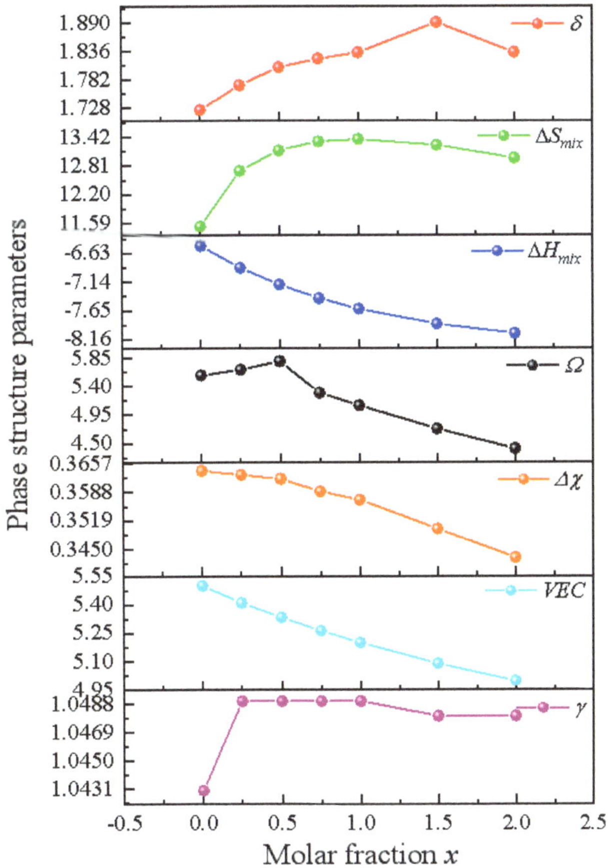

Figure 4. Phase structure parameters of Ti*x*NbMoTaW high-entropy alloy with different Ti contents.

3.1.2. Elastic Properties

Using first-principles density functional theory, the elastic constants of the alloy were calculated. Based on phase structure analysis, the crystal structure of TixNbMoTaW refractory high-entropy alloys (RHEAs) is body-centered cubic (BCC). Due to the symmetry of this structure, there are only three independent elastic constants: C11, C12, and C44. For the mechanical stability of a cubic crystal system, these elastic constants must satisfy the Born–Huang mechanical stability criteria [36–38].

$$C_{ij} > 0(i = 1 \sim 4) \tag{1}$$

$$C_{11} - C_{12} > 0 \tag{2}$$

$$C_{11} + 2C_{12} > 0 \tag{3}$$

As shown in Table 2, the elastic constants of the high-entropy alloy in this study satisfy the mechanical stability criteria, indicating that the alloy is mechanically stable under 0 GPa conditions. To predict the elastic properties of TixNbMoTaW refractory high-entropy alloys (RHEAs) with compositions of x = 0, 0.25, 0.5, 0.75, 1, 1.5, 2, the Voigt–Reuss–Hill (VRH) approximation method was employed to calculate parameters such as the bulk modulus (B), Young's modulus (E), Poisson's ratio (ν), and shear modulus (G) [39,40]. Additionally, the Pugh ratio (B/G), Cauchy pressure (C11–C44), hardness (H), and universal anisotropy index (A^U) [41] were computed using the formulas provided. The results are presented in Table 2.

$$B = B_V = B_R = \frac{c_{11} + 2c_{12}}{3} \tag{4}$$

$$G_R = \frac{5(C_{11} - C_{12})C_{44}}{4C_{44} + 3(C_{11} - C_{12})} \tag{5}$$

$$G_V = \frac{C_{11} - C_{12} + 3C_{44}}{5} \tag{6}$$

$$G = \frac{G_V + G_R}{2} \tag{7}$$

$$E = \frac{9BG}{3B + G} \tag{8}$$

$$v = \frac{3B - 2G}{2(3B + G)} \tag{9}$$

$$H = \frac{(1 - 2v)}{6(1 + v)}E \tag{10}$$

$$A^U = 5\frac{G_V}{G_R} + \frac{B_V}{B_R} - 6 \tag{11}$$

According to references [42,43], adding Ti to NbMoTaW alloys can improve their toughness and ductility. However, the relationship between Ti content and the degree of toughness enhancement remains unclear. In this study, first-principles calculations were used to analyze how varying Ti content affects the degree of toughness improvement in the alloy. The toughness or brittleness of the alloy can be assessed using three parameters: B/G ratio, Poisson's ratio (ν), and Cauchy pressure (C12–C44). The alloy exhibits toughness when it satisfies the conditions B/G > 1.75, ν > 0.26, and C12–C44 > 0; otherwise, it exhibits brittleness [44]. Higher values of these three parameters indicate greater toughness and ductility [44].

Table 2. Calculation results of elastic properties of TixNbMoTaW refractory high-entropy alloy.

Phase	Reference	C_{11}	C_{12}	C_{44}	B	G	E	v	$C_{12}-C_{44}$	B/G	A^u
NbMoTaW	This work	371.24	133.63	63.28	212.83	81.65	244.99	0.37	70.35	2.61	0.49
	Calculation [33]	425	158	86	247	102	270	0.32		2.42	
	Calculation [33]	392	158	79	236	92	245	0.33		2.56	
$Ti_{0.25}$NbMoTW	This work	381.37	143.57	69.71	222.83	86.45	259.38	0.37	73.86	2.58	0.35
$Ti_{0.5}$NbMoTaW	This work	324.15	136.78	73.46	199.24	80.97	242.93	0.36	63.32	2.46	0.10
$Ti_{0.75}$NbMoTW	This work	389.17	153.57	76.87	232.10	91.26	273.78	0.38	76.7	2.54	0.22
TiNbMoTaW	This work	334.43	117.57	66.87	189.85	81.23	243.71	0.35	50	2.33	0.29
$Ti_{1.5}$NbMoTaW	This work	378.17	153.45	71.41	228.35	85.69	257.08	0.37	82.04	2.66	0.25
Ti_2NbMoTaW	This work	392.46	183.15	64.12	252.92	78.10	234.31	0.39	119	3.23	0.29

As the Ti content increases, the bulk modulus (B) generally shows an upward trend, although a slight decrease is observed in the Ti0.5NbMoTaW and TiNbMoTaW alloys. The maximum bulk modulus is found in Ti2NbMoTaW, reaching 252.92 GPa, indicating that the increase in Ti content enhances the alloy's resistance to volumetric deformation. The shear modulus (G) exhibits a fluctuating trend; as the Ti content increases, the G value initially increases, reaching its maximum of 91.26 GPa in Ti0.75NbMoTaW, then decreases to 81.23 GPa in TiNbMoTaW, before slightly recovering to 85.69 GPa in Ti1.5NbMoTaW. The variation in Young's modulus (E) follows a similar trend to that of the shear modulus, increasing initially, then decreasing, and finally rising again. Its maximum value occurs in Ti0.75NbMoTaW, reaching 273.78 GPa. The Pugh ratio increases with the Ti content, indicating improved ductility as the Ti content rises. The maximum Pugh ratio is found in Ti2NbMoTaW, reaching 3.23.

As shown in Figure 5, with increasing Ti content, the B/G ratio, Poisson's ratio (ν), and Cauchy pressure (C12–C44) of TixNbMoTaW high-entropy alloys initially decrease, then increase, with the minimum and maximum values occurring at x = 0.5 and x = 2, respectively. Therefore, it can be concluded that adding a small amount of Ti ($x < 0.5$) to NbMoTaW may slightly reduce its toughness. However, as the Ti content increases, the alloy's toughness improves. At x = 0, the calculated Poisson's ratio is 0.35, which is close to the value of 0.318 reported in the literature [33]. Additionally, at x = 0.75, the alloy not only shows enhanced toughness but also achieves the highest Young's modulus.

The elastic anisotropy index (A^U) reflects the degree of anisotropy in the alloy. A lower A^U value indicates that the alloy exhibits less anisotropy, while a higher A^U value suggests greater anisotropy. When $A^U = 0$, the crystal is isotropic, and the greater the deviation of A^U from 0, the stronger the anisotropy of the material. As shown in Table 2, for TixNbMoTaW high-entropy alloys, when the Ti molar fraction x = 0.5, the A^U coefficient is closest to 0, indicating that the alloy exhibits maximum isotropy.

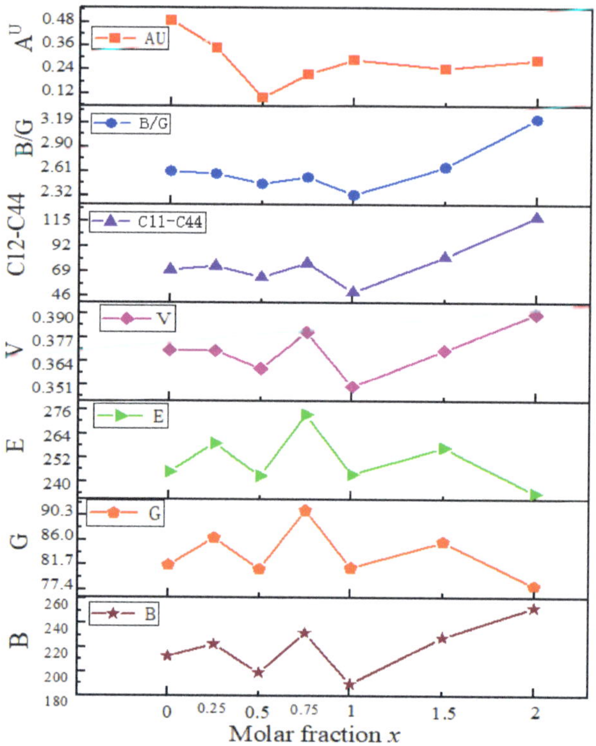

Figure 5. Elastic modulus of TixNbMoTaW high-entropy alloy with different Ti content.

3.1.3. Electronic Structure

Figure 6 illustrates the density of states (DOS) for TixNbMoTaW alloys. The DOS curve remains continuous at the Fermi level (0 eV), with a high density of states, indicating a significant presence of free electrons at the Fermi level, thus exhibiting a distinct metallic characteristic for this alloy system. The lower the electron density at the Fermi level, the more stable the structure is considered to be [45], whereas a higher density of states at this level suggests bonding instability in the system [46]. As shown in Figure 6, the peak DOS below the Fermi level in TixNbMoTaW RHEAs shifts towards higher energy levels as the Ti content in the system increases. In Figure 6, the DOS at the Fermi level first decreases and then continuously increases with increasing Ti content. Therefore, the addition of Ti reduces the structural stability of TixNbMoTaW RHEAs.

The pseudogap in the DOS reflects the covalent character of bonding in the system [47]; a wider pseudogap indicates stronger covalent bonding. In Figure 6, the pseudogap narrows with increasing Ti content, suggesting a decrease in covalent bonding character, an enhancement in metallic bonding, and an improvement in toughness. The higher the electron density, the easier it is to form metallic bonds, which can improve alloy brittleness and enhance toughness [48]. As shown in Figure 6, the electron density near the Fermi level is lowest for $x = 0.25$ and also lower for $x = 0.5$ than for $x = 0$. For $x > 0.5$, andthe DOS increases with higher Ti content. Thus, adding 0.25 and 0.5 molar fractions of Ti to NbMoTaW alloys may actually decrease toughness, while further increases in Ti content would enhance toughness, consistent with the influence of Ti content on alloy brittleness and toughness as indicated by parameters B/G, Poisson's ratio (ν), and the Cauchy pressure (C_{12}~C_{44}).

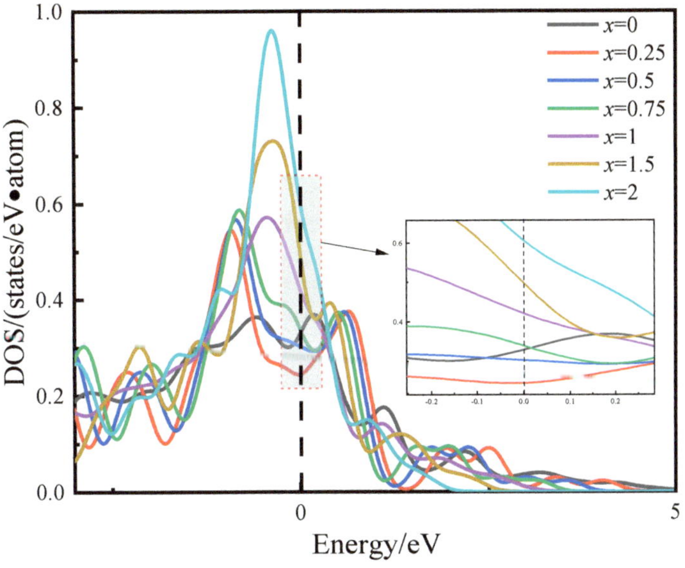

Figure 6. Density of states of Ti*x*NbMoTaW system high-entropy alloys with different Ti contents.

3.2. Experimental Results

3.2.1. Microstructure

To verify the conclusion of a body-centered cubic (BCC) solid solution structure, this section analyzes the microstructure of the alloy through experimental characterization. X-ray diffraction (XRD) was employed to study the TixNbMoTaW refractory high-entropy alloys (HEAs). Figure 7 presents the XRD patterns of the as-cast TixNbMoTaW HEAs. It can be observed that all seven HEAs exhibit a single BCC phase structure, confirming that the experimental results are consistent with the predictions made in the above based on phase formation criteria.

As shown in the figure, all seven high-entropy alloys exhibit four diffraction peaks. The diffraction peaks of the NbMoTaW high-entropy alloy without Ti addition are relatively stronger, with the diffraction peaks at Miller indices of (110) and (200) being significantly more intense than those of the other alloys containing Ti. With the addition of Ti, the diffraction peaks show a tendency to become less pronounced. According to the lattice distortion theory for high-entropy alloys proposed by Yeh J. W. et al. [49], the atomic radius of Ti differs from that of the other elements. The addition of Ti leads to an increase in lattice distortion within the alloy. When subjected to X-ray diffraction, the uneven Bragg planes cause the X-rays to scatter to varying degrees, resulting in a reduction in diffraction intensity and a lowering of the diffraction peaks.

Generally, in XRD patterns, sharper and more intense diffraction peaks indicate a higher crystalline phase fraction and a greater volume proportion of the corresponding crystal structure in the alloy. The analysis of the as-cast TixNbMoTaW alloy XRD patterns in Figure 7 reveals a trend of decreasing stability in the BCC structure as the Ti content increases, suggesting a tendency toward destabilization of the BCC phase. This observation aligns with the predictions based on phase formation parameters and the first-principles calculations on structural stability.

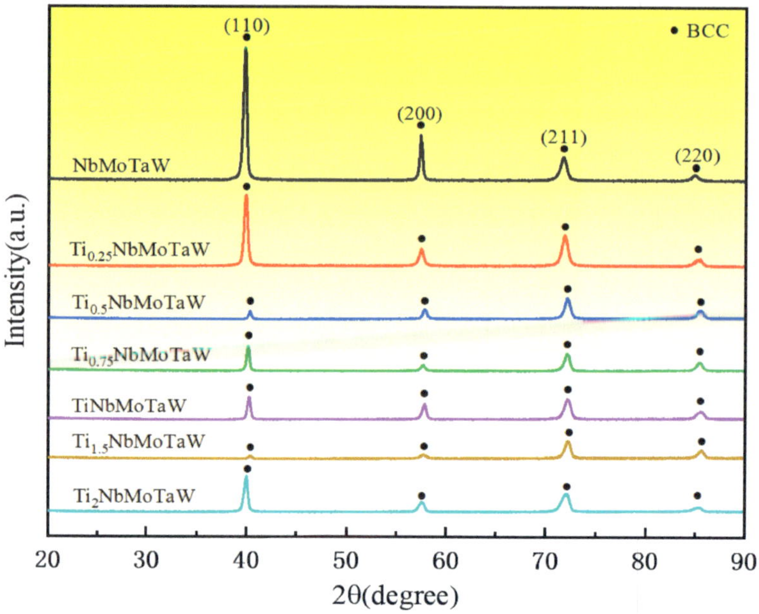

Figure 7. X-ray diffraction patterns of Ti_xNbMoTaW high-entropy alloys in the as-cast state.

Figure 8 presents the backscattered electron (BSE) images of the as-cast TixNbMoTaW refractory high-entropy alloys (HEAs). It can be observed that all the as-cast TixNbMoTaW HEAs exhibit a dendritic structure, and as the Ti content increases, the dendritic structure becomes increasingly dense. The dendritic regions appear darker, while the inter-dendritic areas are brighter. When the Ti molar fractions are 1 and 1.5, there is a noticeable trend of bulging in the dendritic regions, and the surface of the microstructure appears uneven.

Based on the relationship between contrast in backscattered images and atomic number, elements with a higher atomic number generate stronger backscattered electron signals, resulting in a brighter image, while elements with a lower atomic number appear darker. The TixNbMoTaW high-entropy alloys are single-phase BCC solid solutions with no secondary phases present. It can be inferred that different degrees of elemental segregation exist in the seven alloys, with the lower-atomic-number elements, such as Ti, Nb, and Mo, likely concentrating in the darker dendritic regions, while the higher-atomic-number elements, like W and Ta, are likely enriched in the brighter inter-dendritic areas.

The dendritic structure in the alloy consists of dendritic (DR) and interdendritic (ID) regions. The formation of dendrites in TixNbMoTaW high-entropy alloys results from differences in the physical and chemical properties of the elements. During solidification, certain elements become enriched in the dendritic arms, while others concentrate in the interdendritic regions, leading to compositional segregation. To better visualize the degree of segregation, energy-dispersive spectroscopy (EDS) was performed on the seven TixNbMoTaW alloys using scanning electron microscopy (SEM). The analysis included both area and point scans, with the average composition of each element determined from the area scan. Point scans were conducted on five random points in both dendritic and interdendritic regions, excluding the highest and lowest values to determine the specific elemental composition of these regions.

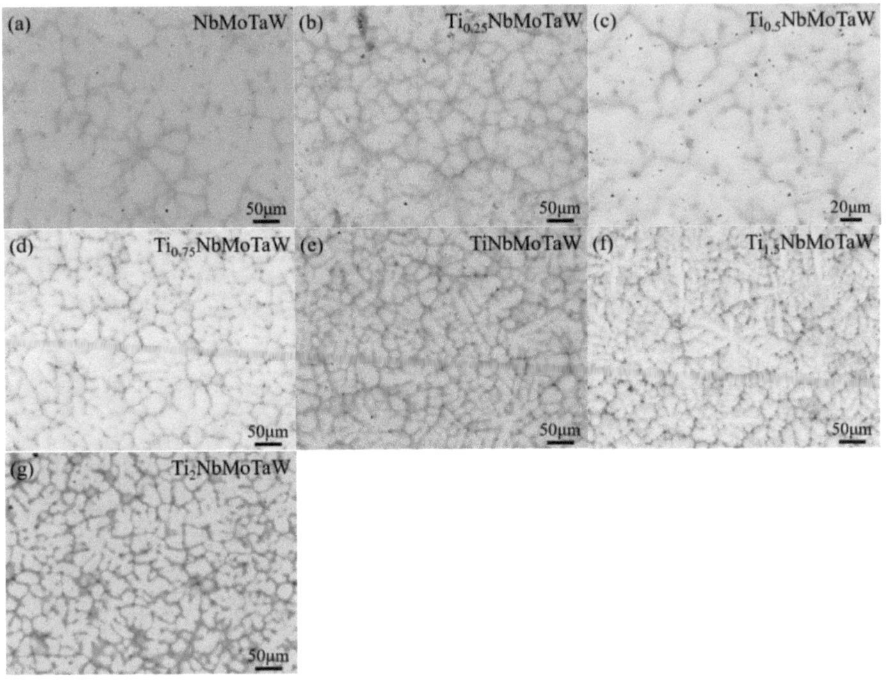

Figure 8. Backscattered electron image of as-cast TixNbMoTaW high-entropy alloys: (**a**) NbMoTaW, (**b**) $Ti_{0.25}$NbMoTaW, (**c**) $Ti_{0.5}$NbMoTaW, (**d**) $Ti_{0.75}$NbMoTaW, (**e**) TiNbMoTaW, (**f**) $Ti_{1.5}$NbMoTaW, (**g**) Ti_2NbMoTaW.

Figure 9 shows the scanning backscattered electron (BSE) images and energy-dispersive spectroscopy (EDS) maps of the seven TixNbMoTaW high-entropy alloys. It can be observed that when the Ti molar fraction in the alloy is greater than or equal to 0.75, the contrast in the EDS maps becomes more distinct. Taking the $Ti_{0.75}$NbMoTaW high-entropy alloy as an example, it is clearly seen that Ti, Nb, and Mo are enriched in the dendritic regions, which appear brighter with more pronounced contrast. In contrast, the interdendritic regions are enriched with Ta and W, and these areas appear darker with lower contrast. By comparing the melting points and atomic masses of the elements in the alloy, it is found that during non-equilibrium solidification, heavy elements with higher melting points and atomic masses, such as W and Ta, tend to accumulate in the interdendritic regions, while lighter elements with lower melting points and smaller atomic masses, such as Ti, Nb, and Mo, are more likely to concentrate in the dendritic regions.

Figure 10 shows the compositional distribution of elements in the dendritic (DR) and interdendritic (ID) regions of the TixNbMoTaW high-entropy alloys. It can be observed that in the dendritic (DR) region, the composition curves of Nb and Mo are higher than those of W and Ta, while in the interdendritic (ID) region, the composition curves of Ta and W consistently lie above those of Nb and Mo. Additionally, the composition curve of Ti reveals that in the dendritic (DR) region, the Ti content increases significantly, reaching approximately 50 at% when $x = 2$, while in the interdendritic (ID) region, the maximum Ti content does not exceed 25 at%. Therefore, Figure 10 visually represents the elemental segregation trends observed in TixNbMoTaW refractory high-entropy alloys.

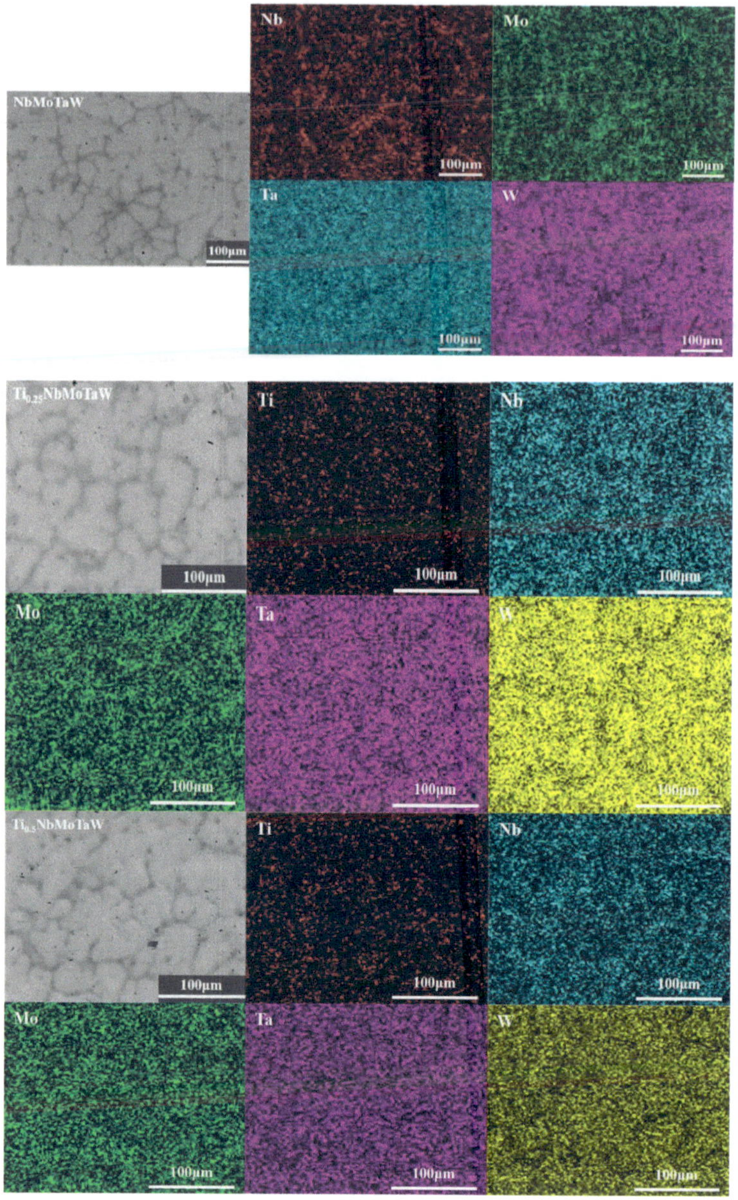

Figure 9. *Cont.*

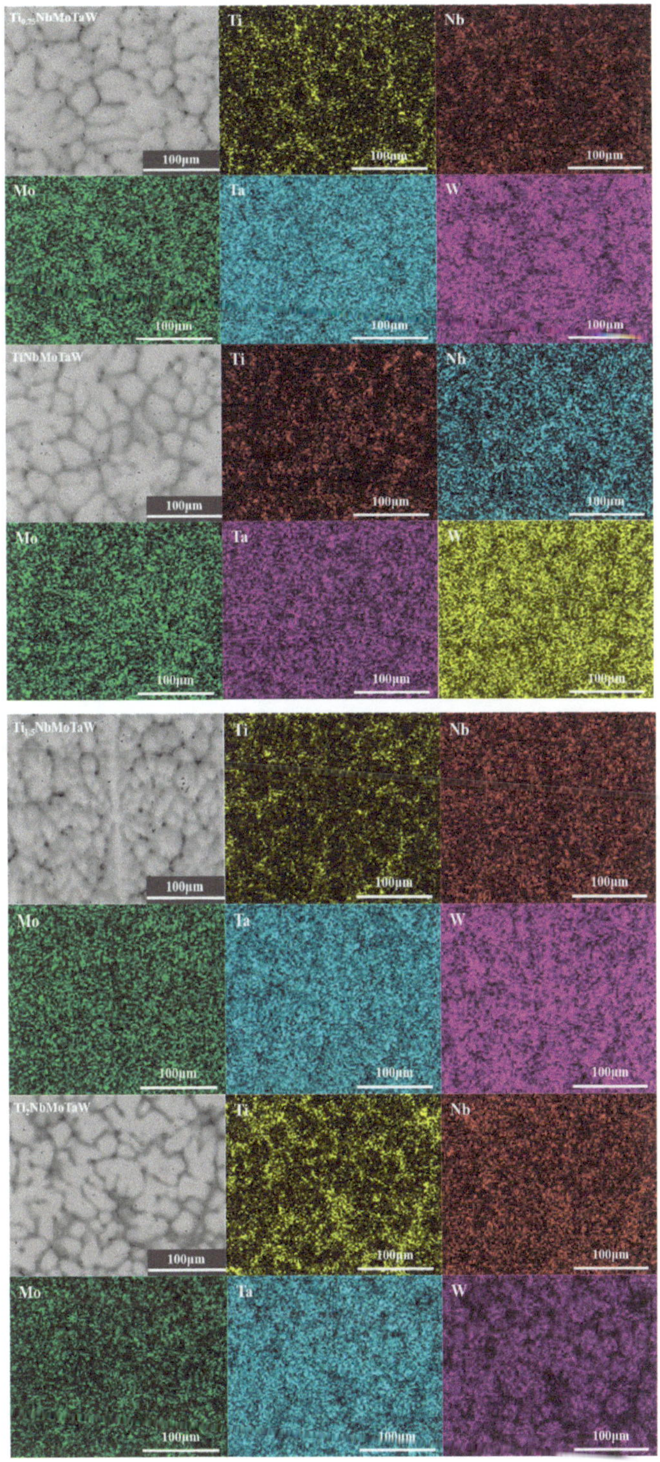

Figure 9. Scanning backscattered electron (SEM) map and energy spectrum of each element (EDS) for Ti_xNbMoTaW high-entropy alloys in cast state.

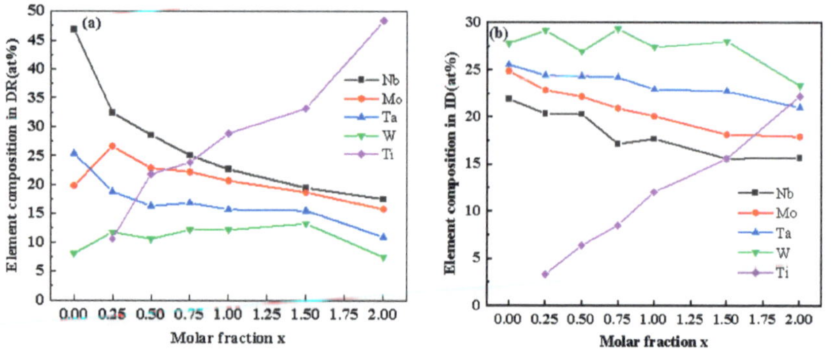

Figure 10. Compositional distribution of each element of $Ti_xNbMoTaW$ high-entropy alloys in dendritic (DR) (**a**) and intergranular (ID) regions (**b**).

Table 3 presents the nominal and actual compositions of each element in the as-cast TixNbMoTaW refractory high-entropy alloys, along with the elemental distribution statistics in the dendritic (DR) and interdendritic (ID) regions. Based on Figures 10 and 11, the compositional segregation trends of the elements in the alloy can be qualitatively analyzed, while Table 3 allows for a quantitative analysis of the degree of segregation for each element. From the data, it can be seen that the actual compositions of the elements in the alloy slightly differ from the nominal compositions, which is due to inevitable losses such as spattering and volatilization during the melting process. The data analysis reveals that the segregation of Ti and W is the most pronounced in the alloy. Ti is significantly enriched in the dendritic regions, while W is predominantly concentrated in the interdendritic regions.

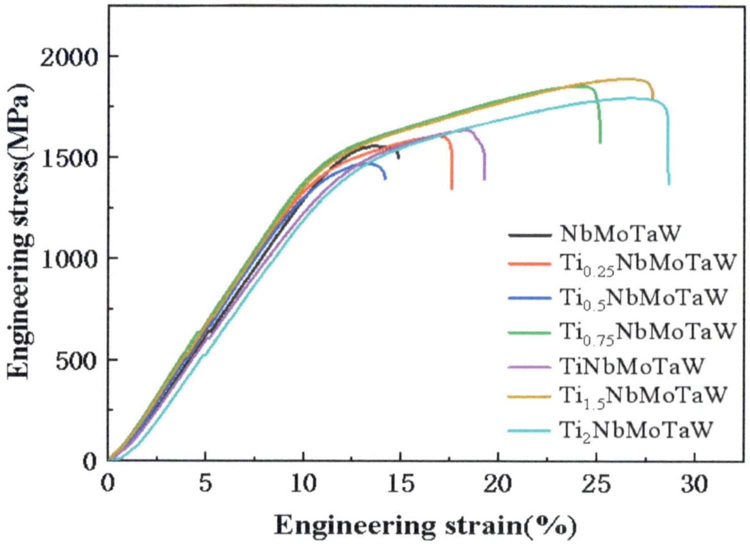

Figure 11. Engineering stress–strain curves for room-temperature compression of refractory high-entropy alloys of the $Ti_xNbMoTaW$ system in cast temper.

The elements Nb and Ta also exhibit some degree of segregation, but compared to Ti and W, they show relatively better uniformity during the alloy solidification process. Lastly, the Mo element shows some degree of segregation only in the NbMoTaW and $Ti_{0.25}$NbMoTaw alloys. However, as the Ti content increases, the composition of Mo in both the dendritic and interdendritic regions, as well as its nominal and actual compositions, become very similar. This indicates that in equimolar or nearly equimolar TixNbMoTaW high-entropy alloys, Mo has the best solubility. Comparing the EDS maps of the five elements in Figure 9, it can also be observed that the contrast for Mo is the least pronounced, with the boundaries between light and dark areas in the image being more blurred. This further confirms that, compared to the other four elements, Mo has better solubility in TixNbMoTaW high-entropy alloys.

Table 3. Compositional distribution of elements in cast Ti_xNbMoTaW high-entropy alloys.

High-Entropy Alloy		Ti	Nb	Mo	Ta	W
	Nominal Composition	0	25.00	25.00	25.00	25.00
	Actual Composition	0	24.39	25.70	24.74	25.17
NbMoTaw	Dendritic Region (DR)	0	46.83	19.80	25.35	8.02
	Interdendritic Region (ID)	0	21.87	24.84	25.51	27.79
	Nominal Composition	5.88	23.53	23.53	23.53	23.53
	Actual Composition	4.57	22.95	23.81	23.30	25.37
$Ti_{0.25}$NbMoTaw	Dendritic Region (DR)	10.52	32.42	26.60	18.75	11.71
	Interdendritic Region (ID)	3.31	20.33	22.82	24.38	29.16
	Nominal Composition	11.11	22.22	22.22	22.22	22.22
	Actual Composition	9.22	22.16	23.04	22.72	22.86
$Ti_{0.5}$NbMoTaw	Dendritic Region (DR)	21.80	28.55	22.82	16.30	10.52
	Interdendritic Region (ID)	6.35	20.28	22.14	24.28	26.95
	Nominal Composition	15.79	21.05	21.05	21.05	21.05
	Actual Composition	15.10	20.63	21.67	21.20	21.40
$Ti_{0.75}$NbMoTaw	Dendritic Region (DR)	23.81	25.04	22.20	16.83	12.12
	Interdendritic Region (ID)	8.47	17.14	20.91	24.15	29.33
	Nominal Composition	20.00	20.00	20.00	20.00	20.00
	Actual Composition	18.37	19.87	20.32	20.43	21.01
Ti_1NbMoTaw	Dendritic Region (DR)	28.84	22.66	20.65	15.72	12.13
	Interdendritic Region (ID)	12.02	17.66	20.06	22.89	27.37
	Nominal Composition	27.27	18.18	18.18	18.18	18.18
	Actual Composition	26.78	18.09	18.29	18.63	18.22
$Ti_{1.5}$NbMoTaw	Dendritic Region (DR)	33.17	19.46	18.66	15.51	13.21
	Interdendritic Region (ID)	15.59	15.59	18.14	22.71	27.97
	Nominal Composition	33.33	16.66	16.66	16.66	16.66
	Actual Composition	31.37	16.41	16.92	17.59	17.72
Ti_2NbMoTaw	Dendritic Region (DR)	48.33	17.54	15.77	10.90	7.46
	Interdendritic Region (ID)	22.18	15.65	17.91	20.97	23.30

3.2.2. Room-Temperature Compression

Figure 11 presents the engineering stress–strain curves of the TixNbMoTaW refractory high-entropy alloys (HEAs) under quasi-static compression at room temperature (25 °C). The strain rate for the uniaxial compression test is set to 5×10^{-4} s^{-1}, and Figure 12 illustrates the compressive plastic strain for each of the seven refractory HEAs. Table 4 summarizes the experimentally measured values for the yield strength, fracture strength, and plastic strain of the TixNbMoTaW HEAs at room temperature.

As shown in Figure 12, along with Table 4, the NbMoTaW HEA exhibits high yield strength, albeit with limited plasticity, with a plastic strain of only 3.8% at room temperature. This finding corroborates the limitations previously noted in Section 1 regarding the NbMoTaW HEA. With increasing Ti content, a slight decrease in alloy yield strength is observed; however, when x > 0.5, both the fracture strength and plastic strain are significantly enhanced. Specifically, Ti0.5NbMoTaW exhibits the lowest values for yield strength, fracture strength, and plastic strain among the tested alloys. At a Ti molar fraction of 0.75, the alloy achieves a maximum yield strength of 1551 MPa, a relatively high fracture strength of 1856 MPa, and a substantially improved plastic strain of 14.6%.

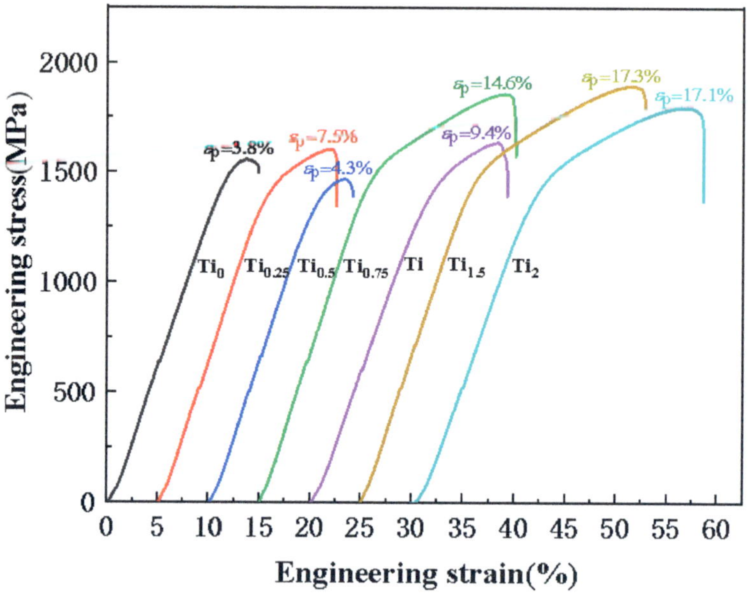

Figure 12. Plastic strain in room-temperature compression of refractory high-entropy alloys of the $Ti_xNbMoTaW$ system in cast temper.

Table 4. Yield strength, fracture strength, and plastic strain of $Ti_xNbMoTaW$ high-entropy alloys at room temperature.

High-Entropy Alloys	Yield Strength (MPa)	Fracture Strength (MPa)	Plastic Strain
NbMoTaW	1510	1558	3.8%
$Ti_{0.25}NbMoTaW$	1443	1605	7.5%
$Ti_{0.5}NbMoTaW$	1430	1470	4.3%
$Ti_{0.75}NbMoTaW$	1551	1856	14.6%
TiNbMoTaW	1467	1636	9.4%
$Ti_{1.5}NbMoTaW$	1506	1893	17.3%
$Ti_2NbMoTaW$	1470	1797	17.1%

When the Ti molar fraction is increased to 1, resulting in an equimolar HEA, the yield strength, fracture strength, and plastic strain of the alloy decrease compared to those of $Ti_{0.75}NbMoTaW$. For Ti contents greater than 1, however, both the yield and fracture strengths of the alloy increase, with a significant improvement in plasticity, as the plastic strain rises from 9.4% at x = 1 to 17.3% and 17.1% at higher Ti contents. The experimental results are consistent with the predictions made in the previous section based on phase formation criteria.

The figures show that as strain increases, the work-hardening rate gradually decreases, which is attributed to the accumulation and interaction of dislocations, leading to increased resistance to dislocation motion. For the Ti = 0 curve, during the elastic stage, the yield stress increases linearly with strain until approximately 10%, after which work hardening begins and the hardening rate gradually slows down as strain increases. When comparing alloys with different Ti contents, Ti = 0.25 exhibits higher stress at the same strain, as the increased Ti content enhances the resistance to dislocation motion. In the case of Ti = 2, due to the significantly higher Ti content, the stress growth rate at lower strains is much greater than for Ti = 0 and Ti = 0.25. This indicates that Ti promotes rapid dislocation accumulation in the early stages of plastic deformation, leading to the formation of more complex dislocation structures and enhancing the material's work-hardening ability. As Ti content increases, the alloy not only exhibits higher stress at the same strain but also accelerates the work-hardening rate in the early stages, suggesting that high-Ti alloys can achieve higher strength more quickly through work hardening during processing.

4. Conclusions

This study addresses the issue of significant room-temperature brittleness in existing NbMoTaW refractory high-entropy alloys (HEAs) by combining first-principles calculations with experimental testing. A series of TixNbMoTaW HEAs (x = 0, 0.25, 0.5, 0.75, 1, 1.5, 2) was designed and developed to investigate the phase formation and mechanical properties. The influence of Ti content on the mechanical properties, particularly plasticity, was systematically studied. The main conclusions are as follows:

(1) The first-principles calculations reveal that TixNbMoTaW alloys exhibit a disordered solid solution phase with a BCC structure. Increasing Ti content lowers the VEC values, enhancing the alloy's toughness and ductility, and favors the formation of the BCC structure.

(2) The bulk modulus (B) increases with Ti content, improving resistance to volumetric deformation. The shear modulus (G) and Young's modulus (E) initially rise, then decrease, before increasing again. Toughness parameters such as the B/G ratio, Poisson's ratio (ν), and Cauchy pressure (C12–C44) improve significantly when the Ti content exceeds 0.75, indicating enhanced plasticity. The density of states analysis shows that increased Ti strengthens metallic bonding and weakens covalent bonding, further confirming improved toughness.

(3) X-ray diffraction (XRD) confirms that all alloys exhibit a single BCC phase, with Ti content reducing BCC stability. The backscattered electron (BSE) images show that the dendritic structure becomes denser with increasing Ti, accompanied by elemental segregation, with Ti and W most prominently segregated.

(4) In quasi-static compression tests at room temperature, the NbMoTaW alloy shows high yield strength but poor plasticity (3.8% strain). As the Ti content increases, yield strength slightly decreases, but fracture strength and plasticity significantly improve when Ti exceeds 0.5.

Author Contributions: R.Z.: Conceptualization, methodology, investigation, simulation, writing—original draft; J.C.: Investigation; J.H.: Investigation; Y.C.: Investigation; Z.Y.: Methodology, investigation; J.L.: Methodology; S.L.: Conceptualization; C.C.: Conceptualization, funding acquisition, resources, supervision, writing—review and editing. All authors have read and agreed to the published version of the manuscript.

Funding: This work was supported by the Regional Collaboration Project of Shanxi Province (No. 202204041101044) and the National Nature Science Foundation of China (52275356).

Institutional Review Board Statement: Not applicable.

Informed Consent Statement: Not applicable.

Data Availability Statement: All data generated or analyzed during this study are included in this published article.

Conflicts of Interest: Yan Cao, Jinhu He, Jianjun Chen, Shiyuan Liu and Zhiqiang Yang are employed by Shanxi Diesel Engine Industry Co., Ltd. The remaining authors declare that the research was conducted in the absence of any commercial or financial relationships that could be construed as a potential conflict of interest.

References

1. Shaheen, M.A.; Foster, A.S.J.; Cunningham, L.S.; Afshan, S. Behaviour of Stainless and High Strength Steel Bolt Assemblies at Elevated Temperatures—A Review. *Fire Saf. J.* **2020**, *113*, 102975. [CrossRef]
2. Wallaert, J.J.; Fisher, J.W. Shear Strength of High-Strength Bolts. *J. Struct. Div.* **1965**, *91*, 99–124. [CrossRef]
3. Može, P.; Beg, D. Investigation of High Strength Steel Connections with Several Bolts in Double Shear. *J. Constr. Steel Res.* **2011**, *67*, 333–347. [CrossRef]
4. Senkov, O.N.; Wilks, G.B.; Miracle, D.B.; Chuang, C.P.; Liaw, P.K. Refractory High-Entropy Alloys. *Intermetallics* **2010**, *18*, 1758–1765. [CrossRef]
5. Chen, B.; Li, X.; Tian, L.; Jia, H.; Li, H.; Li, Y. The CrFeNbTiMox Refractory High-Entropy Alloy Coatings Prepared on the 40Cr by Laser Cladding. *J. Alloys Compd.* **2023**, *966*, 171630. [CrossRef]
6. Cantor, B.; Chang, I.T.H.; Knight, P.; Vincent, A.J.B. Microstructural Development in Equiatomic Multicomponent Alloys. *Mater. Sci. Eng. A* **2004**, *375–377*, 213–218. [CrossRef]
7. Zhang, X.; Cui, X.; Jin, G.; Ding, Q.; Zhang, D.; Wen, X.; Jiang, L.; Wan, S.; Tian, H. Microstructure Evolution and Properties of NiTiCrNbTax Refractory High-Entropy Alloy Coatings with Variable Ta Content. *J. Alloys Compd.* **2022**, *891*, 161756. [CrossRef]
8. Sharma, A. High Entropy Alloy Coatings and Technology. *Coatings* **2021**, *11*, 372. [CrossRef]
9. Nam, S.; Song, Y.; Son, M.; Lee, K.-A.; Yoo, G.H.; Park, E.S.; Choi, H.; Lee, C.S. High-Temperature Mechanical and Thermochemical Properties of NbMoTaW Refractory High-Entropy Alloy Coatings Produced via Nano Particle Deposition System (NPDS). *Int. J. Refract. Met. Hard Mater.* **2021**, *99*, 105594. [CrossRef]
10. Ye, Y.F.; Wang, Q.; Zhao, Y.L.; He, Q.F.; Lu, J.; Yang, Y. Elemental Segregation in Solid-Solution High-Entropy Alloys: Experiments and Modeling. *J. Alloys Compd.* **2016**, *681*, 167–174. [CrossRef]
11. Zhou, Y.J.; Zhang, Y.; Wang, Y.L.; Chen, G.L. Solid Solution Alloys of AlCoCrFeNiTix with Excellent Room-Temperature Mechanical Properties. *Appl. Phys. Lett.* **2007**, *90*, 181904. [CrossRef]
12. Zhu, J.M.; Fu, H.M.; Zhang, H.F.; Wang, A.M.; Li, H.; Hu, Z.Q. Microstructures and Compressive Properties of Multicomponent AlCoCrFeNiMox Alloys. *Mater. Sci. Eng. A* **2010**, *527*, 6975–6979. [CrossRef]
13. Liu, S.; Gao, M.C.; Liaw, P.K.; Zhang, Y. Microstructures and Mechanical Properties of AlxCrFeNiTi0.25 Alloys. *J. Alloys Compd.* **2015**, *619*, 610–615. [CrossRef]
14. Zhao, S.F.; Shao, Y.; Liu, X.; Chen, N.; Ding, H.Y.; Yao, K.F. Pseudo-Quinary Ti20Zr20Hf20Be20(Cu20−xNix) High Entropy Bulk Metallic Glasses with Large Glass Forming Ability. *Mater. Des.* **2015**, *87*, 625–631. [CrossRef]
15. Yang, M.L.; Xu, J.L.; Huang, J.; Zhang, L.W.; Luo, J.M. Wear Resistance of N-Doped CoCrFeNiMn High Entropy Alloy Coating on the Ti-6Al-4V Alloy. *J. Therm. Spray Technol.* **2024**, *33*, 2408–2418. [CrossRef]
16. Kumar, P.; Gou, X.; Cook, D.H.; Payne, M.I.; Morrison, N.J.; Wang, W.; Zhang, M.; Asta, M.; Minor, A.M.; Cao, R.; et al. Degradation of the Mechanical Properties of NbMoTaW Refractory High-Entropy Alloy in Tension. *Acta Mater.* **2024**, *279*, 120297. [CrossRef]
17. Xu, C.; Fang, L.; Xu, G.; Yang, L.; Ouyang, Y.; Tao, X. Mechanical Properties and Oxidation Behavior of NbMoTaWx Refractory High Entropy Alloys. *J. Alloys Compd.* **2024**, *990*, 174390. [CrossRef]
18. Senkov, O.N.; Wilks, G.B.; Scott, J.M.; Miracle, D.B. Mechanical Properties of Nb25Mo25Ta25W25 and V20Nb20Mo20Ta20W20 Refractory High Entropy Alloys. *Intermetallics* **2011**, *19*, 698–706. [CrossRef]
19. Yao, H.W.; Qiao, J.W.; Gao, M.C.; Hawk, J.A.; Ma, S.G.; Zhou, H.F.; Zhang, Y. NbTaV-(Ti,W) Refractory High-Entropy Alloys: Experiments and Modeling. *Mater. Sci. Eng. A* **2016**, *674*, 203–211. [CrossRef]
20. Sheikh, S.; Shafeie, S.; Hu, Q.; Ahlström, J.; Persson, C.; Veselý, J.; Zýka, J.; Klement, U.; Guo, S. Alloy Design for Intrinsically Ductile Refractory High-Entropy Alloys. *J. Appl. Phys.* **2016**, *120*, 164902. [CrossRef]
21. Chan, K.S. A Computational Approach to Designing Ductile Nb-Ti-Cr-Al Solid-Solution Alloys. *Met. Mater. Trans. A* **2001**, *32*, 2475–2487. [CrossRef]
22. Bai, M.; Liu, T.; Liu, B.; Li, Y.; Yu, H.; Zhao, Y.; Yang, C.; Song, L.; Liu, W. Preparation and Properties of Polyurethane Cold Galvanizing Coatings with Phosphoric Acid Modified Zinc Powder. *Surf. Coat. Technol.* **2024**, *489*, 131128. [CrossRef]
23. Rong, J.; Yang, K.; Zhao, H.; Liu, C.; Zhuang, Y.; Tao, S. Tribological Performance of Plasma Sprayed Al2O3–Y2O3 Composite Coatings. *Surf. Coat. Technol.* **2016**, *302*, 487–494. [CrossRef]
24. Qiu, Y.; Thomas, S.; Gibson, M.A.; Fraser, H.L.; Birbilis, N. Corrosion of High Entropy Alloys. *NPJ Mater. Degrad.* **2017**, *1*, 15. [CrossRef]

25. Wang, F.; Liu, Y.; Yu, W.; Jiang, F.; Zarinejad, M.; Tong, Y. Effect of Electropulsing Treatment on the Microstructure and Mechanical Properties of Ti6Al4V Alloy Processed by a Hybrid Method of Directed Energy Deposition and Layer-by-Layer Ultrasonic Impact Peening. *Mater. Sci. Eng. A* **2024**, *908*, 146774. [CrossRef]
26. Qi, L.; Chrzan, D.C. Tuning Ideal Tensile Strengths and Intrinsic Ductility of Bcc Refractory Alloys. *Phys. Rev. Lett.* **2014**, *112*, 115503. [CrossRef]
27. First-Principles Simulation: Ideas, Illustrations and the CASTEP Code—IOPscience. Available online: https://iopscience.iop.org/article/10.1088/0953-8984/14/11/301 (accessed on 28 June 2024).
28. Long, X.; Li, J.; Shen, Z.; Su, Y. Dimensionless Analysis to Determine Elastoplastic Properties of Thin Films by Indentation. *Coatings* **2022**, *12*, 1768. [CrossRef]
29. Long, X.; Lu, C.; Shen, Z.; Su, Y. Identification of Mechanical Properties of Thin-Film Elastoplastic Materials by Machine Learning. *Acta Mech. Solida Sin.* **2023**, *36*, 13–21. [CrossRef]
30. Long, X.; Li, Y.; Shen, Z.; Su, Y.; Gu, T.; Siow, K.S. Review of uniqueness challenge in inverse analysis of nanoindentation. *J. Manuf. Process.* **2024**, *131*, 1897–1916. [CrossRef]
31. Perdew, J.P. Generalized Gradient Approximation Made Simple. *Phys. Rev. Lett.* **1996**, *77*, 3865–3868. [CrossRef]
32. Payne, M.C. Iterative Minimization Techniques for Ab Initio Total-Energy Calculations: Molecular Dynamics and Conjugate Gradients. *Rev. Mod. Phys.* **1992**, *64*, 1045–1097. [CrossRef]
33. Hu, Y.L.; Bai, L.H.; Tong, Y.G.; Deng, D.Y.; Liang, X.B.; Zhang, J.; Li, Y.J.; Chen, Y.X. First-Principle Calculation Investigation of NbMoTaW Based Refractory High Entropy Alloys. *J. Alloys Compd.* **2020**, *827*, 153963. [CrossRef]
34. Senkov, O.N.; Scott, J.M.; Senkova, S.V.; Miracle, D.B.; Woodward, C.F. Microstructure and Room Temperature Properties of a High-Entropy TaNbHfZrTi Alloy. *J. Alloys Compd.* **2011**, *509*, 6043–6048. [CrossRef]
35. Chen, H.; Kauffmann, A.; Gorr, B.; Schliephake, D.; Seemüller, C.; Wagner, J.N.; Christ, H.-J.; Heilmaier, M. Microstructure and Mechanical Properties at Elevated Temperatures of a New Al-Containing Refractory High-Entropy Alloy Nb-Mo-Cr-Ti-Al. *J. Alloys Compd.* **2016**, *661*, 206–215. [CrossRef]
36. Goumri-Said, S.; Kanoun, M.B. Theoretical investigations of structural, elastic, electronic and thermal properties of Damiaoite PtIn2. *Comput. Mater. Sci.* **2008**, *43*, 243–250. [CrossRef]
37. Beckstein, O.; Klepeis, J.E.; Hart, G.L.W.; Pankratov, O. First-Principles Elastic Constants and Electronic Structure of α-Pt2Si and c. *Phys. Rev. B* **2001**, *63*, 134112. [CrossRef]
38. Noheda, B.; Gonzalo, J.A.; Cross, L.E.; Guo, R.; Park, S.-E.; Cox, D.E.; Shirane, G. Tetragonal-to-Monoclinic Phase Transition in a Ferroelectric Perovskite: The Structure of PbZr0.52Ti0.48O3. *Phys. Rev. B* **2000**, *61*, 8687–8695. [CrossRef]
39. Korozlu, N.; Colakoglu, K.; Deligoz, E.; Aydin, S. The Elastic and Mechanical Properties of MB12 (M = Zr, Hf, Y, Lu) as a Function of Pressure. *J. Alloys Compd.* **2013**, *546*, 157–164. [CrossRef]
40. Pugh, S.F. XCII. Relations between the Elastic Moduli and the Plastic Properties of Polycrystalline Pure Metals. *Lond. Edinb. Dublin Philos. Mag. J. Sci.* **1954**, *45*, 823–843. [CrossRef]
41. Bi, X.; Hu, X.; Li, Q. Effect of Co Addition into Ni Film on Shear Strength of Solder/Ni/Cu System: Experimental and Theoretical Investigations. *Mater. Sci. Eng. A* **2020**, *788*, 139589. [CrossRef]
42. Feng, X.B.; Zhang, J.Y.; Wang, Y.Q.; Hou, Z.Q.; Wu, K.; Liu, G.; Sun, J. Size Effects on the Mechanical Properties of Nanocrystalline NbMoTaW Refractory High Entropy Alloy Thin Films. *Int. J. Plast.* **2017**, *95*, 264–277. [CrossRef]
43. Shao, L.; Yang, M.; Ma, L.; Tang, B.-Y. Effect of Ta and Ti Content on High Temperature Elasticity of HfNbZrTa1−xTix Refractory High-Entropy Alloys. *Int. J. Refract. Met. Hard Mater.* **2021**, *95*, 105451. [CrossRef]
44. Mouhat, F.; Coudert, F.-X. Necessary and Sufficient Elastic Stability Conditions in Various Crystal Systems. *Phys. Rev. B* **2014**, *90*, 224104. [CrossRef]
45. Tian, J.; Zhao, Y.; Wen, Z.; Hou, H.; Han, P. Physical Properties and Debye Temperature of Al7Cu2Fe Alloy under Various Pressures Analyzed by First-Principles. *Solid State Commun.* **2017**, *257*, 6–10. [CrossRef]
46. Carlsson, A.E.; Meschter, P.J. Relative Stability of LI2, DO22, and DO23 Structures in MAl3 Compounds. *J. Mater. Res.* **1989**, *4*, 1060–1063. [CrossRef]
47. Gschneidner, K.A.; Ji, M.; Wang, C.Z.; Ho, K.M.; Russell, A.M.; Mudryk, Y.; Becker, A.T.; Larson, J.L. Influence of the Electronic Structure on the Ductile Behavior of B2 CsCl-Type AB Intermetallics. *Acta Mater.* **2009**, *57*, 5876–5881. [CrossRef]
48. Gao, F.; Qu, J. Elastic Moduli of (Ni,Cu)3Sn4 Ternary Alloys from First-Principles Calculations. *J. Electron. Mater.* **2010**, *39*, 2429–2434. [CrossRef]
49. Yeh, J.-W.; Chang, S.-Y.; Hong, Y.-D.; Chen, S.-K.; Lin, S.-J. Anomalous Decrease in X-Ray Diffraction Intensities of Cu–Ni–Al–Co–Cr–Fe–Si Alloy Systems with Multi-Principal Elements. *Mater. Chem. Phys.* **2007**, *103*, 41–46. [CrossRef]

coatings

Article

Toughened Bamboo-Fiber-Modified Epoxy Resin: A Novel Polymer Coating for Superior Interfacial Compatibility

Baoping Zou [1,2], Kai Huang [1,2] and Jingyuan Ma [1,2,*]

1 School of Civil Engineering and Architecture, Zhejiang University of Science and Technology, Hangzhou 310023, China; zoubp@zust.edu.cn (B.Z.); honore2@163.com (K.H.)
2 Zhejiang-Singapore Joint Laboratory for Urban Renewal and Future City, Hangzhou 310023, China
* Correspondence: majingyuan@zust.edu.cn; Tel.: +86-13269660086

Abstract: Epoxy resin is regarded as a reliable option for coating advanced materials owing to its outstanding strength, adhesion, and stability. However, its relatively weak toughness compared to common materials has limited its application. In this study, the toughness of epoxy resin was enhanced by incorporating bamboo fibers, and a novel polymer coating material for bamboo-fiber-reinforced epoxy resin was developed. Different fiber pretreatment methods were employed to address the issue of poor interfacial performance between bamboo fibers and epoxy resin, aiming to optimize its performance as an advanced material coating. The effects of curing agents, fiber mesh sizes, fiber contents, and fiber pretreatment methods on the mechanical properties of the fiber-modified resin composites were investigated. The findings indicate that the JH45 and T31 curing agents were more effective in promoting the homogeneous dispersion of fibers within the epoxy resin. Additionally, bamboo fibers modified with KH550 exhibited enhanced interfacial properties: the tensile strength of the composite demonstrated a respective increase of 31.1% and 27.0% compared to untreated fibers. Increasing the mesh size proved advantageous for improving tensile properties, albeit potentially impacting the compressive properties. Particularly noteworthy was the significantly enhanced interfacial compatibility between bamboo fibers treated with the silane coupling agent KH550 and the epoxy resin. Analysis using Fourier-transform infrared spectroscopy (FTIR), scanning electron microscopy (SEM), and X-ray diffraction (XRD) revealed that treating bamboo fibers with sodium hydroxide effectively enhanced bonding at the fiber–resin interface. This enhancement was attributed to the combined effects of bamboo fiber hydrolysis and delamination reactions. The silane coupling agent promoted the chemical reaction between bamboo fibers and epoxy resin through grafting, thereby strengthening the cross-linking property of the composites. These findings offer valuable insights into the design and fabrication of natural-fiber-reinforced polymer composites suitable for coating advanced materials.

Keywords: natural fiber; epoxy resin; polymer; electronic packaging; coating

Academic Editor: Jean-François Berret

Received: 9 January 2025
Revised: 30 January 2025
Accepted: 3 February 2025
Published: 5 February 2025

Citation: Zou, B.; Huang, K.; Ma, J. Toughened Bamboo-Fiber-Modified Epoxy Resin: A Novel Polymer Coating for Superior Interfacial Compatibility. *Coatings* **2025**, *15*, 181. https://doi.org/10.3390/coatings15020181

1. Introduction

With the increasing precision of circuit component protection in the field of electronic packaging, fiber-reinforced epoxy polymer composites (FRPs) can effectively protect electronic components from external physical shocks and vibrations due to their excellent mechanical properties and chemical resistance and good adhesive properties, avoiding damage to the internal structures of the components. As an excellent coating material, they are widely used for chip encapsulation, the protection of sensitive components, and structural protection in aerospace engineering [1–5]. Among fiber-reinforced epoxy resin materials, inorganic synthetic fibers, such as glass fibers and carbon fibers, have limited applications due to their high cost and adverse effects on the environment, despite having significant advantages in terms of improving the strength and stiffness of composites [6–9]. The brittleness of glass fibers and the electrical conductivity of carbon fibers make them unsuitable in the field of electronic coatings, especially for protection against electromagnetic interference or cost-sensitive applications. In contrast, bamboo fiber, as a natural plant fiber [10], not only possesses good mechanical properties and a high specific surface area [11] but also has the advantages of natural abundance and environmental degradability. Combining bamboo fiber with epoxy resin can not only effectively improve the crack resistance, toughness, and durability of composites but also reduce the brittleness of epoxy resin [12,13], thus expanding the scope of its application in coating materials.

However, poor interfacial compatibility between fibers and epoxy resins is a key bottleneck limiting the performance of composites [14]. Dong [15] analyzed the effect of fiber interfacial roughness on polymer cross-linking properties. Yazman [16] added multi-walled carbon nanotubes (MWCNTs) and alumina (Al_2O_3) nanoparticles (NPs) to improve the epoxy surface, Prasanthi [17] nano-strengthened carbon fibers, and Kim [18] improved the fiber surface to improve the overall mechanical properties through covalent bonding and introducing silica, among other ways. While bamboo fibers are a natural organic material, their surface contains a large amount of hydrophilic and phenolic hydroxyls [19], and as the surface of the epoxy resin is hydrophobic, this hydrophilic–hydrophobic mismatch leads to poor interfacial bonding between the two, which, in turn, affects the mechanical properties and durability of the composites. To solve this problem, many researchers adopted fiber surface modification techniques: Huang [20] et al. treated the fiber surface via alkali immersion and used resin transfer molding (RTM), Ding [21] investigated the effect of different oxidation systems on the surface of bamboo fibers, Bai [22] used a fiber wetting agent to increase the tensile strength by 99.7%, and Wang [23], as well as Li [24], used silane modification and nanosilica filling to improve the interfacial adhesion between fibers and epoxy resin, thus optimizing the properties of the composites.

Most current research focuses on the application of larger-diameter bamboo fibers [25–27], which perform well in improving the mechanical properties and crack resistance of composites. However, in practical encapsulation engineering, large-size fibers cannot uniformly fill tiny gaps and spaces, which can affect the electrical performance of the encapsulation and affect signal transmission, especially for miniaturized and highly integrated electronic components at the micrometer level [28,29], where the required fiber sizes tend to be smaller. During the high-precision electronic encapsulation process, the size, dispersion, and treatment of fibers significantly impact the material's properties, such as its fluidity, curing time, and interfacial bonding [30]. The compatibility between fillers and epoxy resins is a crucial factor influencing material properties [31]. Finer bamboo powder fibers can achieve better integration with the epoxy resin matrix and preserve the original fibers' properties [32]. Their increased specific surface area due to the smaller fiber diameter facilitates improved interactions with epoxy resin-based materials, making them ideal for coating and filling fine voids. The aim of this study is to investigate how to improve the interfacial compatibility between bamboo fibers and epoxy resin through different chemical modification treatments of bamboo powder; to introduce a silane coupling agent to strengthen the interfacial properties between the fibers and epoxy [33]; and to regulate the curing agent, fiber mesh size, fiber content, etc., to prepare a composite coating material with excellent performance in electronic encapsulation. Considering the unique requirements for the curing temperature, setting time, fluidity, and fiber size of the material, particularly in encapsulation engineering, this study will conduct in-depth analysis in this area to promote the application of epoxy resin matrix composites in electronic engineering coatings.

2. Materials and Methods

2.1. Materials

In this study, considering the fillability of the grouting material for voids and inhomogeneous areas within the electronic package structure, it is noted that excessively large fiber fillers may affect packaging performance, while the morphology of bamboo fibers with small diameters would be weakened. Bamboo powder with different particle sizes was used as a natural-fiber-reinforced polymer, specifically 100, 200, 300, and 500-mesh bamboo powder. The raw material was from Guangdong moso bamboo fibers. The base material for the epoxy resin (E51) is an ordinary bisphenol A-type liquid product produced by Hangzhou Wuhuigang Adhesive Co., Ltd., with an epoxy equivalence of 185–200 g/eq and an epoxy value of 0.48–0.54 eq/mg. To compare the performance of the grouting material, four specific curing agents were selected, all of which are produced by Hangzhou Wuhuigang Adhesive Co., Ltd. These curing agents are T31, JH45, JH593, and 650 polyamide. (a) The T31 epoxy resin curing agent is a phenol-formaldehyde amine compound containing hydroxyl, amine, and secondary groups as well as other reactive groups. (b) The JH593 epoxy resin curing agent is a low-viscosity curing agent synthesized by the reaction between epoxy compounds and fatty amines. (c) JH-45 is a new, excellent epoxy resin curing agent containing special amines and phenol, formaldehyde, etc. It is modified via the Mannich condensation reaction. The molecular structure contains phenol hydroxyl (-OH), primary amine (-NH_2), and secondary amine (-NH-) condensates. (d) The 650 polyamide resin curing agent is a class of polymer compound obtained by the reaction of diamine and dibasic acid. Additionally, during the treatment of bamboo fibers, NaOH (1 mol/L) was used as an alkaline treatment to remove impurities from the fiber surface. The silane coupling agent KH550 (γ-aminopropyltriethoxysilane; density, 0.946 g/cm^3; boiling point, 217 °C; and purity, $\geq$97%) was used to improve the surface properties of the fibers. Anhydrous ethanol (99.7%) was employed to hydrolyze KH550, and nano-SiO_2 powder modified with KH550 (particle size, 20 nm; purity, 99 wt%) was used to further enhance the surface properties of the fibers.

2.2. Preparation of Bamboo-Fiber-Modified Epoxy Resin

Generally speaking, epoxy resin has a hydrophobic surface, while natural plant fibers contain a large number of free hydroxyl and phenolic hydroxyl groups, which exhibit strong hydrophilic properties. This leads to poor interfacial compatibility between the two and significantly affects the properties of composites. Chemically modifying the surface of natural plant fibers can effectively improve their interfacial compatibility with epoxy resin. Four differently modified types of bamboo powder fibers were generated according to different treatments, as shown in Figure 1: (a) untreated fiber (BF); (b) NaOH-treated fiber (N-BF), for which dry bamboo fibers were placed in a 5% NaOH solution, magnetically stirred (SN-MS-1, Lichen Bangxi Instrument (Changzhou) Co., Ltd., Changzhou, China) at room temperature for 6 h, and then washed with deionized water to neutralize them and dried at 80 °C to obtain alkali-treated bamboo fibers; (c) KH550-treated fibers (KH550-BF), a 5% silane coupling agent solution, using an ethanol solution (mass ratio of anhydrous ethanol to water = 9:1) as the solvent, was prepared by stirring continuously for 1 h to achieve complete hydrolyzation, and then, alkali-treated bamboo fibers (5%) were added, and the solution was stirred magnetically at 60–80 °C for 4 h, filtered, and dried at 60–80 °C to obtain silane coupling agent-modified bamboo fiber; (d) nano-SiO_2-treated fibers (SiO_2-BF) [34], which were obtained by adding 3% KH550 surface-modified silica nanopowder during the magnetic stirring stage in (c).

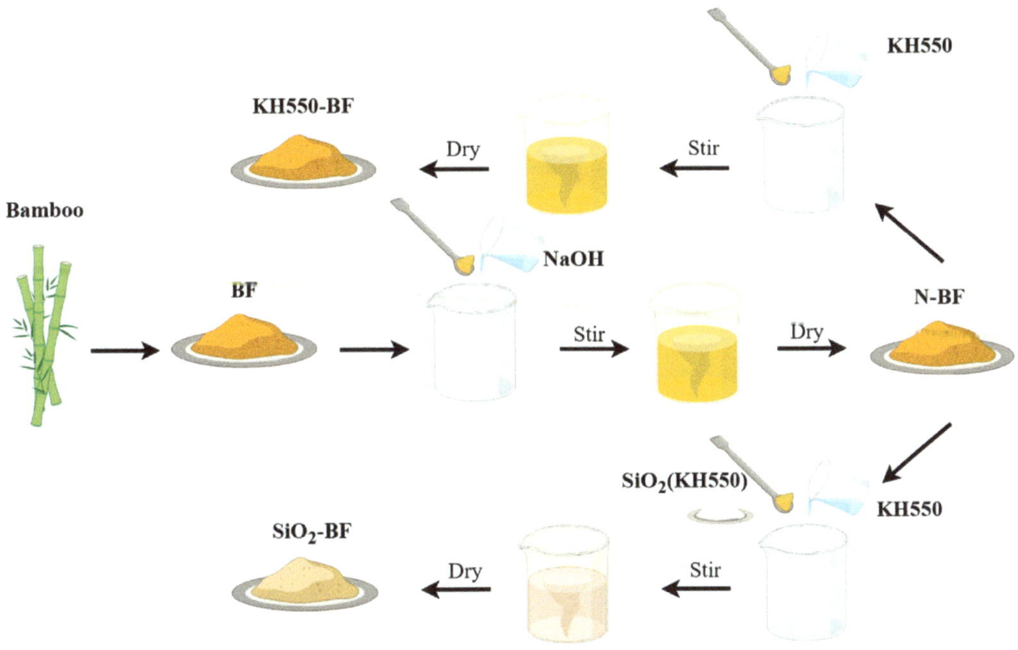

Figure 1. Methods of making bamboo–epoxy-resin-based materials with different treatments.

To investigate the various bamboo–epoxy resin natural polymer grouting materials employed in electronic packaging, four key factors were examined: curing agent, fiber mesh count, fiber content, and fiber treatment technique. Four variations in these factors were studied, as outlined in Table 1, to fabricate samples and assess these factors' influence on the overall material characteristics.

Table 1. Variables of bamboo–epoxy-resin-based materials.

Curing Agent	Fiber Content	Mesh Number of Fibers	Fiber Handling
JH45	5%	100-mesh	(a)
JH593	10%	200-mesh	(b)
T31	15%	300-mesh	(c)
650 polyamide	20%	500-mesh	(d)

2.3. *Performance Testing*

2.3.1. Gel Time

Gel time was tested using the inverted cup method, where the surface of the mixture did not show any obvious flow or collapse within 5 s. A physical probe (PA125, Shenzhen Huarong fa electronic testing Co., Ltd., Shenzhen, China) was also used to measure the gel state of the epoxy resin, and the difficulty of probe insertion was increased to the point where it could not easily penetrate the mixture to determine the gel time.

2.3.2. Viscosity

The viscosity and operation time of the composites under different curing agents were recorded by using an NDJ-1-type rotational viscometer (Lichen Bangxi Instrument (Changzhou) Co., Ltd., Changzhou, China). The epoxy resin must meet specific viscosity criteria to ensure smooth flow during electronic packaging. A viscosity exceeding

1000 mPa·s would hinder the efficient dispensing of the slurry onto electronic components. Hence, the moment when viscosity hits 1000 mPa·s is considered the operation time.

2.3.3. Mechanical Properties

According to the GB/T 2567-2021 standard [35], tensile and compressive specimens were created using a universal material testing machine CMT-410 (Shenzhen new think material testing Co., Ltd., Shenzhen, China) (Figure 2) to test the effect of fiber reinforcement on the composite material. The test parameters were a loading rate of 2 mm/min and a sensor load range of 1 kN. The effect of different fiber modification treatments on the performance of epoxy resin-based composites was analyzed to optimize the resin formula and the processing technology.

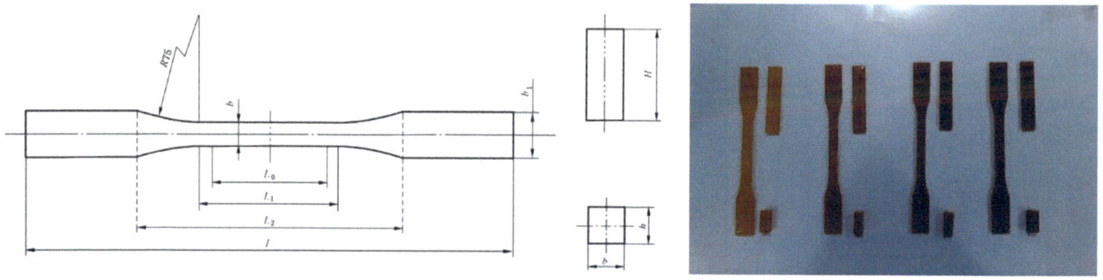

(a) (b) (c)

Figure 2. GB/T 2567-2021 compressive and tensile test block standard: (**a**) tensile test block; (**b**) compressive test block; and (**c**) actual test block.

2.4. Microproperty Characterization

2.4.1. Infrared Spectroscopy (FTIR)

Fourier infrared spectra of different modified fibers were recorded by using a Thermo Fisher Scientific Nicolet iS20 spectrometer (Thermo Fisher Scientific, Waltham, MA, USA), Fifty milligrams were placed in a mold and kept at a pressure of 20 MPa using a hydraulic press. The test parameters were a spectral range of 400–4000 cm^{-1}, a resolution of 4 cm^{-1}, and a scanning number of 32 to further analyze the surface functional groups of bamboo fibers before and after modification.

2.4.2. X-Ray Diffraction (XRD)

XRD analysis of the composite samples was carried out by using a Rigaku Ultima IV (Rigaku Corporation, Tokyo, Japan) from Japan to determine the major element contents on the surface of the bamboo fibers before and after modification. The test parameters were Cu targets, a tube voltage of 40 kV, a tube current of 40 mA, and a scanning angle of 5–40°.

2.4.3. Scanning Electron Microscopy (SEM)

SEM was performed by using a Hitachi SUI5IO scanning electron microscope (Hitachi Production Co., Ltd., Tokyo, Japan), and the dried samples were coated on or adhered to a tape of about 0.5 cm × 0.5 cm and metal-sprayed for 15 min. The microscopic morphology of the surface of the bamboo fibers and the cross-section of the bamboo-fiber-reinforced epoxy composites before and after modification were observed to assess the interfacial bond strength of the fibers and the resin.

3. Results

3.1. Microscopic Characterization Under Different Fiber Treatments

3.1.1. Scanning Electron Microscopy Analysis

SEM images of the surface of bamboo fibers before and after modification are shown in Figure 3. As shown by comparisons with the untreated bamboo fibers, the impurities and non-cellulosic substances on the fiber surface were removed by treating the natural bamboo powder with sodium hydroxide, which enhanced the surface roughness of the fibers and thus improved the affinity of the fibers and the resin bonding. As for the silane coupling agent-treated bamboo fibers, the surface roughness was further enhanced and some particles were adhered to the fiber surface, which showed an obvious layered structure. In contrast, the roughness of SiO_2-treated fibers was not significantly enhanced, although particles were adhered to the surface.

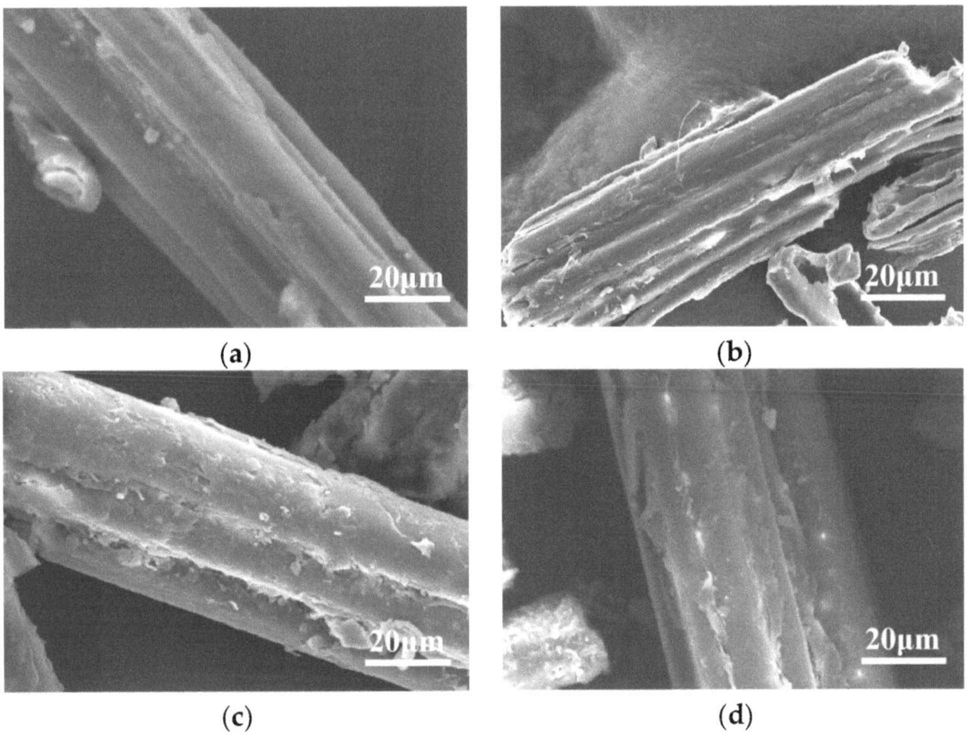

Figure 3. SEM images of the bamboo fiber surface before and after modification. (**a**) BF; (**b**) N-BF; (**c**) KH550-BF; and (**d**) SiO_2-BF.

Figure 4 shows the SEM images of bamboo–epoxy resin matrix composite sections. Observing the bamboo-fiber-reinforced epoxy resin composite cross-section, the untreated bamboo fibers did not produce an actual connection with the epoxy resin due to the presence of interfacial repulsion. By treating the bamboo fibers with sodium hydroxide and KH550, due to the enhancement of roughness and the improvement of surface hydroxyl groups, polymerization of the epoxy resin matrix composites improved. A large amount of epoxy resin was cross-linked in the cross-section, which improves the affinity of the fibers and the bonding force of the resin.

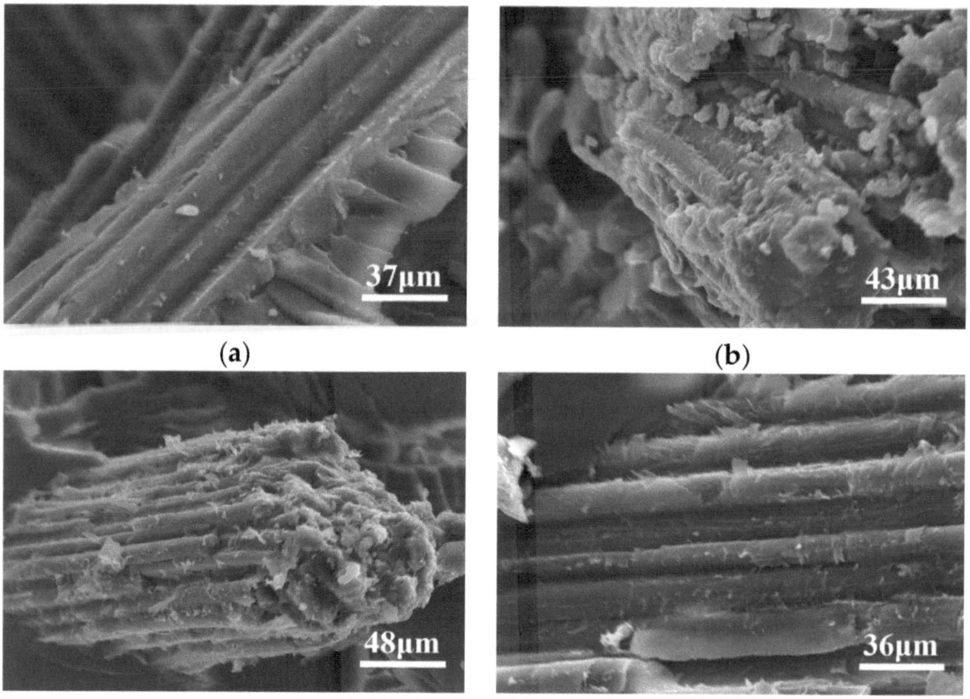

(a) (b) (c) (d)

Figure 4. SEM images of bamboo–epoxy resin matrix composite sections. (**a**) BF; (**b**) N-BF; (**c**) KH550-BF; and (**d**) SiO_2-BF.

3.1.2. Analysis of Functional Groups Under Different Fiber Treatments

The IR spectra of bamboo fibers under different treatments are shown in Figure 5, and the corresponding characteristic peaks are listed in Table 2. The results of IR spectral analysis showed that the unmodified fibers had characteristic peaks at 3362 cm^{-1} and 2930 cm^{-1}, which were related to O-H and C-H stretching vibrations, respectively, in addition to the characteristic peaks of C=O, C=C, C-H, and C-O. After sodium hydroxide treatment, the O-H peak at 3362 cm^{-1} was broadened, and the C=O peak at 1753 cm^{-1} was red-shifted to 1651 cm^{-1}, indicating that sodium hydroxide promoted the formation of inter-fiber hydrogen bonding. The intensity of the C=C vibration at 1590 cm^{-1} was reduced, and the C-O peak at 1249 cm^{-1} disappeared, which indicated that the alkali treatment destroyed the fiber skeleton and depolymerized some of the structures. Characteristic Si-O peaks appeared at 667 and 581 cm^{-1} in the silane coupling agent-treated fibers, and the O-H peak at 3362 cm^{-1} and Si-O-Si peak at 1031 cm^{-1} were enhanced, which indicated that the silane coupling agent was effectively bonded on the surface of the fibers. After nanosilica treatment, new characteristic peaks appeared in the low-frequency region, and the peak intensity at 1110 cm^{-1} increased and the half-peak width narrowed, indicating that nanosilica modification resulted in a more ordered fiber structure. According to the chemical formula shown in Figure 6, alkali treatment exposed more hydroxyl groups and improved the chemical bonding between the fiber and the resin. The interfacial bonding between the fiber and resin is improved by using a silane coupling agent to form a chemically bonded layer on the fiber surface. The silane coupling agent KH550 typically has a hydrophilic group and a chemical group that reacts with the epoxy resin matrix to effectively bridge the interface between the natural fiber and the epoxy resin.

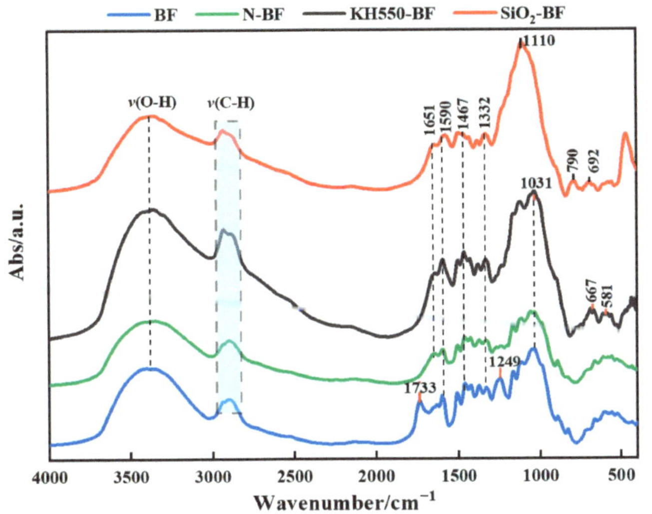

Figure 5. Infrared spectra of fibers under different treatments.

Table 2. Characteristic peaks of fibers under different treatments.

Functional Groups / Treatment	O-H	C-H	C=O	C-C	C-O	Si-O
BF	3362 cm^{-1}	2930 cm^{-1}	1733 cm^{-1}	1590 cm^{-1}	1249 cm^{-1}	-
N-BF	3362 cm^{-1}	2930 cm^{-1}	1651 cm^{-1}	1590 cm^{-1}	1249 cm^{-1}	-
KH550-BF	3362 cm^{-1}	2930 cm^{-1}	-	-	1031 cm^{-1}	667, 581 cm^{-1}
SiO_2-BF	3362 cm^{-1}	2930 cm^{-1}	-	-	1100 cm^{-1}	790, 692 cm^{-1}

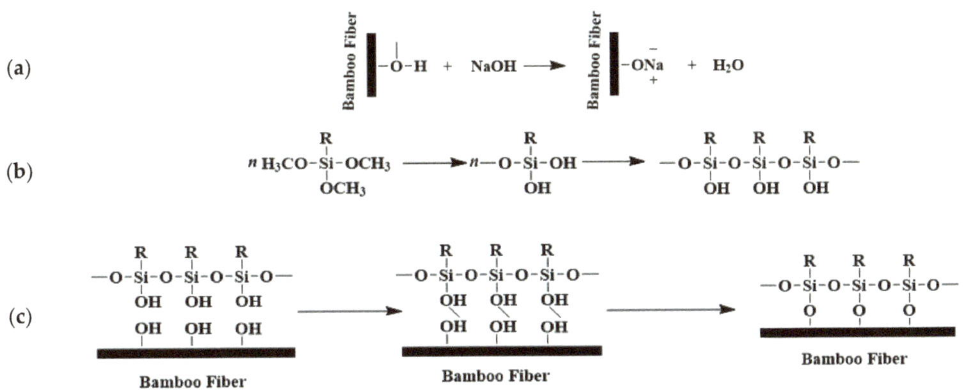

Figure 6. Reaction formula of bamboo fiber and silane coupling agent treatment under alkali treatment. (**a**) Reaction formula of bamboo fiber with NaOH; (**b**) reaction formula of KH550 hydrolysis; and (**c**) reaction formula of bamboo fiber with KH550.

3.1.3. Physical Phase Analysis with Different Fiber Treatments

As shown in the XRD plot in Figure 7, the bamboo fibers mainly contain two C phases, corresponding to the hexagonal phases PDF#50-0927 and PDF#49-0944. After alkali treatment, the peak attributed to PDF#50-0927 near 26.5° decreases, indicating that this crystalline surface is actually etched, and the peak near 29.6° is enhanced, indicating that it is favorable for this crystalline surface to be exposed. Additionally, the peak near 16.2° is enhanced for graphitic carbon, suggesting that the alkali treatment weakened the graphitization of bamboo fibers. This treatment mainly changed the surface structure of the bamboo fiber, reduced the stability of the crystal surface structure, weakened the degree of graphitization, and improved the flexibility as well as surface affinity of the fibers. For the bamboo fiber treated with the silane coupling agent, the peak near 26.5° disappeared and the characteristic graphitic carbon peak near 16.2° weakened. The graphitization degree was higher, and a sharp peak also appeared near 24.0° due to PDF#50-0927, which indicated that the increase in C phases promoted an increase in the crystalline phase and graphitization degree of bamboo fibers. This shows that the silane coupling agent helps to improve the bonding of the fiber surface with other materials, as well as the performance of the composites. The nanosilica-treated bamboo fibers were similar to alkali-treated bamboo fibers, with the characteristic peak near 26.5° disappearing and the graphitic carbon characteristic peak near 16.2° increasing. This shows that the silane coupling agent's main role is to enhance the surface structure of the fiber and make the fiber more ordered.

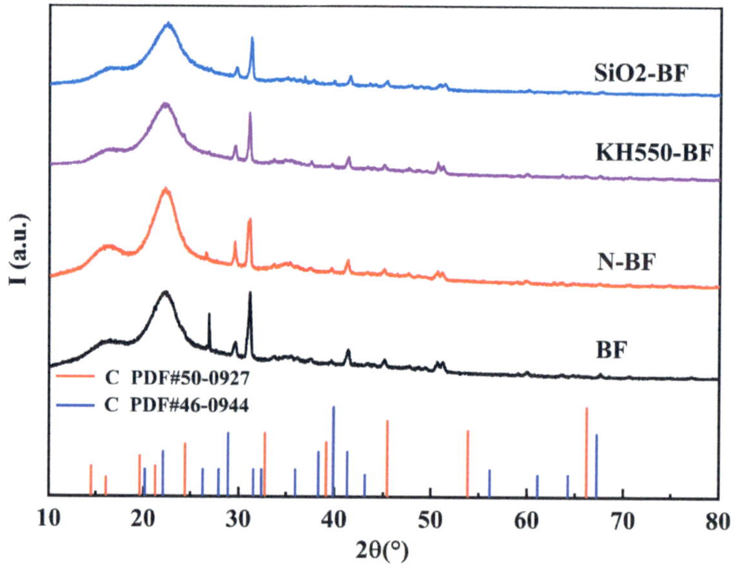

Figure 7. XRD plots of bamboo fiber under different treatments.

3.2. *Effect of Curing Agents on the Mechanical Properties of Bamboo–Epoxy-Resin-Based Materials*

The curing agent is a key factor in determining the final properties of epoxy resins [36]. An ideal electronic packaging material should possess adequate strength and hardness to endure mechanical pressure, impact vibration forces, and external forces over extended periods of use. This endurance is crucial to maintaining the structural stability of the encapsulated area, ensuring the protection of electronic components and, at the same time, ensuring that the slurry can fill internal voids and inhomogeneous areas in the electronic component's structure and that the viscosity and solidification time are suitable for encapsulation. In this study, due to the addition of natural fibers, it is also necessary to ensure that the natural fibers are uniformly distributed in the epoxy resin matrix. The compressive strength (Figure 8a) and tensile strength (Figure 8b) of the fiber/epoxy composites cured with the addition of 5% bamboo fibers (100-mesh) with different curing agents were tested. The compressive strength of the epoxy system based on the T31 curing agent was 110.7 MPa and the tensile strength was 47.4 Mpa; the compressive strength of the epoxy system based on the JH45 curing agent was 90.5 Mpa and the tensile strength was 35.6 Mpa; the compressive strength of the epoxy system based on the JH593 curing agent was 73.4 Mpa and the tensile strength was 37.7 Mpa; and the compressive strength of the epoxy system based on the 650 polyamide curing agent was 17.1 Mpa and the tensile strength was 51.9 Mpa. High cross-linking densities were still reached after the addition of bamboo fibers with the use of aliphatic amine curing agents containing amino groups (T31, JH45, and JH593), and they were able to increase the hardness, strength, and heat resistance of the resin [37]. Hydroxyl groups can form hydrogen bonds, and this cross-linking mode usually results in a better toughness of the cured epoxy resin. In contrast, polyacid-based (low-molecular-weight 650 polyamides) and acid-based curing agents lead to lower cross-linking densities compared to amine-based curing agents, resulting in resins with a lower strength but higher toughness.

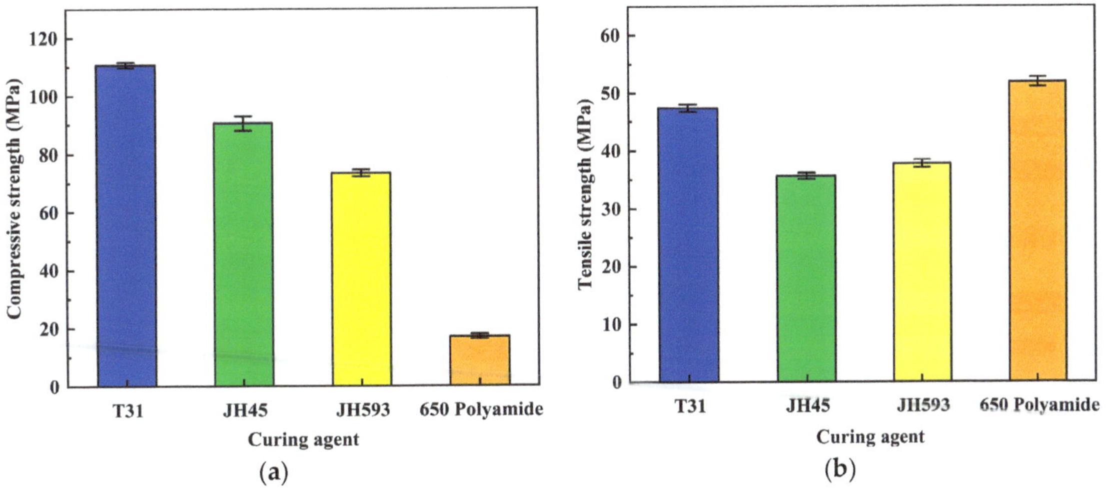

Figure 8. Mechanical properties of bamboo–epoxy-resin-based materials with different curing agents. (**a**) Compressive strength; (**b**) tensile strength.

Meanwhile, the viscosity of the mixed slurry obtained from the curing process with the addition of JH593 and low-molecular-weight 650 polyamide is low, as shown in Table 3, which is not conducive to the homogeneous distribution of bamboo fibers in the composites. In addition, the operable time of JH593 is short, while the curing time of low-molecular-weight 650 polyamide at room temperature (25 °C) is too long and the compressive strength after curing is low, which makes it unsuitable for practical packaging engineering applications.

Table 3. Curing agent properties.

Curing Agent	Amine Value (mgKOH/g)	Viscosity (mPa·s)	Solidification Time (25 °C, min)	Operable Time (25 °C, min)	Ratio of Curing Agent to E51
JH45	500 ± 20	500 ± 50	40	25	1:3.3
JH593	550 ± 50	100 ± 50	20	14	1:4
T31	480 ± 20	600 ± 50	120	40	1:3.5
650 polyamide	220 ± 20	50 ± 10	1200	90	1:1

3.3. *Effect of Fiber Mesh on the Mechanical Properties of Bamboo–Epoxy-Resin-Based Materials*

The effect of fiber mesh on the mechanical properties of the epoxy resin was analyzed using the same curing agent and uniformly treated fibers. As shown in Figure 9a, at 10% bamboo fiber content, the compressive strengths of 200, 300, and 500-mesh bamboo fibers decreased by 1.1%, 0.8%, and 4.4%, respectively, compared with 100-mesh bamboo fibers, indicating that changing the mesh number in the range of 100–500-mesh has a limited effect on the overall compressive properties of the composites. However, as shown in Figure 9b, the 500-mesh bamboo fibers exhibited the highest tensile strength (39.8 MPa) at a content of 10%, whereas the 200-mesh bamboo fibers had the weakest tensile strength (35.9 Mpa) at the same content, which is a reduction in tensile properties of 10.8%. This indicates that with an equal amount of fibers, for higher-mesh bamboo fibers, a larger specific surface area is in contact with the epoxy resin, and they are able to transfer stress more efficiently.

3.4. *Effect of Fiber Content on the Mechanical Properties of Bamboo–Epoxy-Resin-Based Materials*

An analysis of the effect of fiber content on the mechanical properties of the epoxy resin (Figure 9) shows that the compressive strength of the bamboo–epoxy-resin-based material increases and then decreases with the increase in bamboo fiber content, reaching a peak at a content of 10%. However, the change in the compressive strength is relatively small, which suggests that changing the bamboo fiber content in the range of 5%~20% has a limited influence on the overall compressive properties of the composites. The tensile strength also increased with the increase in bamboo fiber content, reaching a peak at 10%. This is consistent with the trend in compressive strength, which further verifies the enhancement of the composites' mechanical properties with the appropriate bamboo fiber content. This indicates that reasonable fiber content can effectively enhance the performance of composites with the same mesh number, while too much or too little bamboo powder particles lead to weakened cross-linking between the fibers and the epoxy resin, which affects the densification of the material and thus reduces the compressive strength.

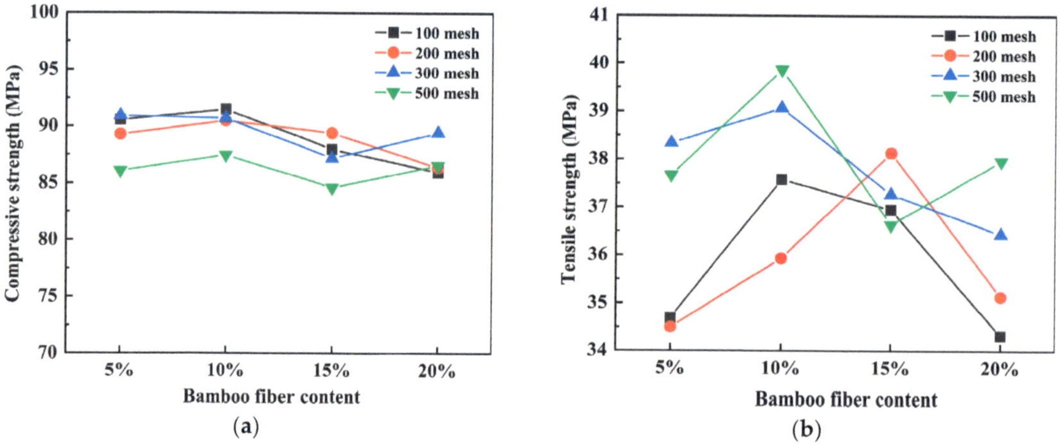

Figure 9. Mechanical properties of bamboo–epoxy resin-based materials with different fiber contents and mesh numbers. (**a**) Compressive strength and (**b**) tensile strength.

3.5. *Effect of Fiber Treatment on the Mechanical Properties of Bamboo–Epoxy-Resin-Based Materials*

Four kinds of bamboo powder fibers with different fiber treatment methods were studied to verify the effect of fiber mesh number on the mechanical properties of composites. As shown in Figure 10, compared with the untreated fibers in Section 3.3, the tensile strength of composites with 10% bamboo fiber content increased by 10.9% to 41.7 MPa when alkali-treated bamboo powder was used as the curing agent. The tensile strength of composites with silane coupling agent KH550-treated bamboo powder increased by 31.1% to 49.3 Mpa. However, the SiO_2-treated bamboo powder shows only a weak enhancement in tensile strength compared to the untreated fibers at 10% and 20% content. When T31 was used as the curing agent, the tensile strength of composites formed by alkali-treated bamboo powder increased by 6.4% to 54.7 Mpa compared with untreated fibers, and the tensile strength of composites formed by the silane coupling agent KH550 increased by 27.0% to 65.3 Mpa, while the tensile strength of bamboo powder treated by SiO_2 generally decreased. The results show that the surface roughness of alkali-treated bamboo powder increased the contact area with the epoxy resin matrix and the silane coupling agent enhanced the surface roughness of bamboo powder and grafted epoxy resin, both of which enhanced the degree of cross-linking between fiber and epoxy resin.

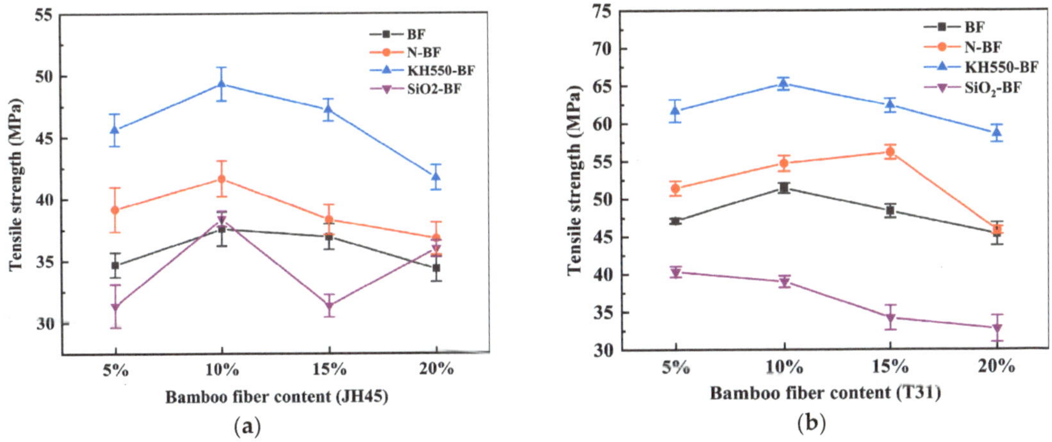

Figure 10. Tensile strength of bamboo–epoxy-resin-based materials with different fiber treatments. (**a**) JH45 curing agent; (**b**) T31 curing agent.

4. Conclusions

In this study on the development of environmentally friendly epoxy resin encapsulation coatings, bamboo powder fibers with different chemical modifications, achieved using different curing agents, fiber webs, and fiber contents, were investigated. The conclusions of this study are summarized as follows: (1) Selecting a curing agent is of great significance to electronic encapsulation engineering, and JH45 and T31 can ensure the uniform distribution of bamboo fibers in natural polymer composites as well as infer excellent mechanical and encapsulation properties. (2) The mesh number and fiber content of bamboo powder also affect the performance of composites. Bamboo fibers with a mesh of 500 improved the tensile properties by 10.8%, whereas a bamboo fiber content of 10% resulted in the best tensile properties. (3) The basic mechanical properties of the bamboo fibers under different treatments were tested, and the surface functional groups, material composition, and micro-morphology of the fibers were analyzed using Fourier-transform infrared spectroscopy, X-ray diffraction, and scanning electron microscopy. Alkali treatment could expose more hydroxyl groups in natural bamboo fibers, increase the surface roughness, and effectively improve the tensile strength by 10.9% compared to the composite material formed by untreated bamboo fibers combined with epoxy resin. Silane coupling agents can form chemical groups while altering the surface roughness of bamboo fibers, and by attaching tiny particles to the epoxy resin matrix for a chemical reaction, they also significantly enhance the cross-linking effect between bamboo fibers and epoxy resin. The obtained composite material shows a 31.1% increase in tensile strength compared to the composite with untreated fibers. In conclusion, bamboo powder, as a green material, shows great potential in encapsulated coating applications. This study solved the key problem of poor interfacial bonding between the polymer matrix and bamboo fibers, providing new insights and paving the way for the development of new high-performance polymer/fiber composite coatings.

Author Contributions: Conceptualization, B.Z.; methodology, J.M.; validation, J.M. and K.H.; formal analysis, J.M. and K.H.; investigation, B.Z. and K.H.; data curation, J.M. and K.H.; writing—original draft preparation, B.Z. and K.H.; writing—review and editing, J.M. and K.H.; funding acquisition, B.Z. All authors have read and agreed to the published version of the manuscript.

Funding: This research was funded by the National Natural Science Foundation of China (grant number: 42477185), the Zhejiang Provincial Natural Science Foundation of China (grant number: LQN25E080017), the Zhejiang Provincial Department of Education Research Project (grant number: Y202351878), the Zhejiang University of Science and Technology Youth Science Foundation Project (grant number: 2023QN087), the Zhejiang Province University Students Science and Technology Innovation Program (grant number: 0201310P28), and the Zhejiang University of Science and Technology Graduate Research and Innovation Fund (grant number: 2023yjskc10).

Institutional Review Board Statement: Not applicable.

Informed Consent Statement: Not applicable.

Data Availability Statement: Data are contained within the article.

Acknowledgments: The authors acknowledge the School of Civil Engineering, Zhejiang University of Science and Technology and Zhejiang-Singapore Joint Laboratory for Urban Renewal and Future City for the experimental space provided.

Conflicts of Interest: The authors declare no conflicts of interest.

References

1. Alim, M.A.; Abdullah, M.Z.; Aziz, M.S.A.; Kamarudin, R.; Gunnasegaran, P. Recent advances on thermally conductive adhesive in electronic packaging: A review. *Polymers* **2021**, *13*, 3337. [CrossRef] [PubMed]
2. Aparna, A.; Sethulekshmi, A.S.; Saritha, A.; Joseph, K. Recent advances in superhydrophobic epoxy based nanocomposite coatings and their applications. *Prog. Org. Coat.* **2022**, *166*, 106819. [CrossRef]
3. Zhao, Y.; Xu, R.; Xiao, Y.; Wang, H.; Zhang, W.; Zhang, G. Mechanical performances of phenolic modified epoxy resins at room and high temperatures. *Coatings* **2022**, *12*, 643. [CrossRef]
4. Sanchez, W.A.L.; Li, J.; Chiu, H.; Cheng, C.; Chiou, K.; Lee, T.; Chiu, C.W. Highly thermally conductive epoxy composites with aln/bn hybrid filler as underfill encapsulation material for electronic packaging. *Polymers* **2022**, *14*, 2950. [CrossRef] [PubMed]
5. Zhang, F.; Zhang, J.; Zhang, K.; Zhong, X.; He, M.; Qiu, H.; Gu, J. Highly thermally conductive liquid crystalline epoxy resin vitrimers with reconfigurable, shape-memory, photo-thermal, and closed-loop recycling performance. *Adv. Sci.* **2024**, *12*, e2410362. [CrossRef]
6. Ogbonna, V.E.; Popoola, A.P.; Popoola, O.M.; Adeosun, S.O. A review on corrosion, mechanical, and electrical properties of glass fiber-reinforced epoxy composites for high-voltage insulator core rod applications: Challenges and recommendations. *Polym. Bull.* **2022**, *79*, 6857–6884. [CrossRef]
7. Hua, Y.; Li, F.; Hu, N.; Fu, S. Frictional characteristics of graphene oxide-modified continuous glass fiber reinforced epoxy composite. *Compos. Sci. Technol.* **2022**, *223*, 109446. [CrossRef]
8. Sheinbaum, M.; Sheinbaum, L.; Weizman, O.; Dodiuk, H.; Kenig, S. Toughening and enhancing mechanical and thermal properties of adhesives and glass-fiber reinforced epoxy composites by brominated epoxy. *Compos. Part B-Eng.* **2019**, *165*, 604–612. [CrossRef]
9. George, J.S.; Vijayan, P.P.; Paduvilan, J.K.; Salim, N.; Sunarso, J.; Kalarikkal, N.; Hameed, N.; Thomas, S. Advances and future outlook in epoxy/graphene composites for anticorrosive applications. *Prog. Org. Coat.* **2022**, *162*, 106571. [CrossRef]
10. Mousavi, S.R.; Zamani, M.H.; Estaji, S.; Tayouri, M.I.; Arjmand, M.; Jafari, S.H.; Nouranian, S.; Khonakdar, H.A. Mechanical properties of bamboo fiber-reinforced polymer composites: A review of recent case studies. *J. Mater. Sci.* **2022**, *57*, 3143–3167. [CrossRef]

11. Karimah, A.; Ridho, M.R.; Munawar, S.S.; Adi, D.S.; Ismadi; Damayanti, R.; Subiyanto, B.; Fatriasari, W.; Fudholi, A. A review on natural fibers for development of eco-friendly bio-composite: Characteristics, and utilizations. *J. Mater. Res. Technol.* **2021**, *13*, 2442–2458. [CrossRef]
12. Hu, H.; Cheng, X.; Liu, Z.; Chen, X.; Piao, R.; Xiang, C. Influence of Bamboo Fibers Weight Fraction on the Quasi-Static and Dynamic Compressive Responses of Epoxy Matrix Composites. *J. Nat. Fibers* **2023**, *20*, 2246646. [CrossRef]
13. Alsuwait, R.B.; Souiyah, M.; Momohjimoh, I.; Ganiyu, S.A.; Bakare, A.O. Recent development in the processing, properties, and applications of epoxy-based natural fiber polymer biocomposites. *Polymers* **2023**, *15*, 145. [CrossRef] [PubMed]
14. Lu, T.; Jiang, M.; Jiang, Z.; Hui, D.; Wang, Z.; Zhou, Z. Effect of surface modification of bamboo cellulose fibers on mechanical properties of cellulose/epoxy composites. *Compos. Part B Eng.* **2013**, *51*, 28–34. [CrossRef]
15. Dong, J.; Jia, C.; Wang, M.; Fang, X.; Wei, H.; Xie, H.; Zhang, T.; He, J.; Jiang, Z.; Huang, Y. Improved mechanical properties of carbon fiber-reinforced epoxy composites by growing carbon black on carbon fiber surface. *Compos. Sci. Technol.* **2017**, *149*, 75–80. [CrossRef]
16. Yazman, S.; Uyaner, M.; Karabork, F.; Akdemir, A. Effects of nano reinforcing/matrix interaction on chemical, thermal and mechanical properties of epoxy nanocomposites. *J. Compos. Mater.* **2021**, *55*, 4257–4272. [CrossRef]
17. Prasanthi, P.P.; Kumar, M.S.R.N.; Chowdary, M.S.; Madhav, V.V.V.; Saxena, K.K.; Mohammed, K.A.; Khan, M.I.; Upadhyay, G.; Eldin, S.M.; Kumar, G. Mechanical properties of carbon fiber reinforced with carbon nanotubes and graphene filled epoxy composites: Experimental and numerical investigations. *Mater. Res. Express.* **2023**, *10*, 025308. [CrossRef]
18. Kim, D.; Chung, I.; Kim, G. Study on mechanical and thermal properties of fiber-reinforced epoxy/hybrid-silica composite. *Fibers Polym.* **2013**, *14*, 2141–2147. [CrossRef]
19. Chen, X.; Wang, X.; Luo, X.; Chen, L.; Li, Y.; Xu, J.; Liu, Z.; Dai, C.; Miao, H.; Liu, H. Bamboo as a naturally-optimized fiber-reinforced composite: Interfacial mechanical properties and failure mechanisms. *Compos. Part B Eng.* **2024**, *279*, 111458. [CrossRef]
20. Huang, J.; Young, W. The mechanical, hygral, and interfacial strength of continuous bamboo fiber reinforced epoxy composites. *Compos. Part B-Eng.* **2019**, *166*, 272–283. [CrossRef]
21. Ding, R.; Liu, X.; Yu, H.; Shi, S.Q.; Han, G.; Cheng, W. Effects of different oxidation systems on the interfacial properties of bamboo fiber/epoxy resin composites. *Surf. Interfaces* **2024**, *45*, 103843. [CrossRef]
22. Bai, T.; Wang, D.; Yan, J.; Cheng, W.; Cheng, H.; Shi, S.Q.; Wang, G.; Han, G. Wetting mechanism and interfacial bonding performance of bamboo fiber reinforced epoxy resin composites. *Compos. Sci. Technol.* **2021**, *213*, 108951. [CrossRef]
23. Wang, D.; Bai, T.; Cheng, W.; Xu, C.; Wang, G.; Cheng, H.; Han, G. Surface modification of bamboo fibers to enhance the interfacial adhesion of epoxy resin-based composites prepared by resin transfer molding. *Polymers* **2019**, *11*, 2107. [CrossRef] [PubMed]
24. Li, X.; Li, G.; Su, X. Nanosio2 strengthens and toughens epoxy resin/basalt fiber composites by acting as a nano-mediator. *J. Polym. Eng.* **2018**, *39*, 10–15. [CrossRef]
25. Xie, J.; Qi, J.; Hu, T.; De Hoop, C.F.; Hse, C.Y.; Shupe, T.F. Effect of fabricated density and bamboo species on physical-mechanical properties of bamboo fiber bundle reinforced composites. *J. Mater. Sci.* **2016**, *51*, 7480–7490. [CrossRef]
26. Oliveira, M.; Neves, V.; Banea, M.D. Mechanical and thermal characterization of bamboo and interlaminar hybrid bamboo/synthetic fibre-reinforced epoxy composites. *Materials* **2024**, *17*, 1777. [CrossRef]
27. Yu, Y.; Huang, Y.; Zhang, Y.; Liu, R.; Meng, F.; Yu, W. The reinforcing mechanism of mechanical properties of bamboo fiber bundle-reinforced composites. *Polym. Compos.* **2019**, *40*, 1463–1472. [CrossRef]
28. Otiaba, K.C.; Bhatti, R.S.; Ekere, N.N.; Mallik, S.; Alam, M.O.; Amalu, E.H.; Ekpu, M. Numerical study on thermal impacts of different void patterns on performance of chip-scale packaged power device. *Microelectron Reliab.* **2012**, *52*, 1409–1419. [CrossRef]
29. Zou, B.; Yin, J.; Liu, Z.; Long, X. Transient rock breaking characteristics by successive impact of shield disc cutters under confining pressure conditions. *Tunn. Undergr. Space Technol.* **2024**, *150*, 105861. [CrossRef]
30. Lakshmaiya, N.; Kaliappan, S.; Patil, P.P.; Ganesan, V.; Dhanraj, J.A.; Sirisamphanwong, C.; Wongwuttanasatian, T.; Chowdhury, S.; Channumsin, S.; Channumsin, M.; et al. Influence of oil palm nano filler on interlaminar shear and dynamic mechanical properties of flax/epoxy-based hybrid nanocomposites under cryogenic condition. *Coatings* **2022**, *12*, 1675. [CrossRef]
31. Kostenko, M.; Stetsyshyn, Y.; Harhay, K.; Melnyk, Y.; Donchak, V.; Gubriy, Z.; Kracalik, M. Impact of the functionalized clay nanofillers on the properties of the recycled polyethylene terephthalate nanocomposites. *J. Appl. Polym. Sci.* **2024**, *141*, e55543. [CrossRef]
32. Dutta, H.; Gurusamy, P.; Halder, S. Enhancement of thermal and mechanical properties of polymer composite using a novel milled bamboo filler. *Polym. Compos.* **2023**, *44*, 8731–8749. [CrossRef]
33. Jin, B.H.; Jang, J.; Kang, D.J.; Yoon, S.; Im, H. Epoxy-based siloxane composites for electronic packaging: Effect of composition and molecular structure of siloxane matrix on their properties. *Compos. Sci. Technol.* **2022**, *224*, 109456. [CrossRef]
34. Luo, L.; Ma, Q.; Wang, Q.; Ding, L.; Gong, Z.; Jiang, W. Study of a nano-sio 2 microsphere-modified basalt flake epoxy resin coating. *Coatings* **2019**, *9*, 154. [CrossRef]
35. *GB/T 2567-2021*; Resin Casting Body Performance Test Method. China National Standards: Shenzhen, China, 2021.

36. Aziz, T.; Haq, F.; Farid, A.; Cheng, L.; Chuah, L.F.; Bokhari, A.; Mubashir, M.; Tang, D.Y.Y.; Show, P.L. The epoxy resin system: Function and role of curing agents. *Carbon Lett.* **2024**, *34*, 883. [CrossRef]
37. Garcia, F.G.; Leyva, M.E.; de Queiroz, A.A.A.; Simoes, A.Z. Durability of adhesives based on different epoxy/aliphatic amine networks. *Int. J. Adhes. Adhes.* **2011**, *31*, 177–181. [CrossRef]

MDPI AG
Grosspeteranlage 5
4052 Basel
Switzerland
Tel.: +41 61 683 77 34

Coatings Editorial Office
E-mail: coatings@mdpi.com
www.mdpi.com/journal/coatings

www.ingramcontent.com/pod-product-compliance
Lightning Source LLC
Chambersburg PA
CBHW041344210926
51850CB00013B/2990
* 9 7 8 3 7 2 5 8 3 7 6 5 6 *